Maritime Watchstanding Plans:
Origins, Variants and Effectiveness

James C. Miller, Ph.D.

Copyright 2015 James C. Miller ebook (Smashwords) with color figures

Copyright 2018 James C. Miller paperback (CreateSpace)

Series: Shiftwork, Fatigue and Safety

Book 1. *Shiftwork: An Annotated Bibliography* (Smashwords,

Book 2. *Fundamentals of Shiftwork Scheduling, 3rd Edition: Fixing Stupid* (Smashwords, 2015, ebook. CreateSpace, 2017, paperback)

Book 3. *Anatomy of a Fatigue-Related Accident* (Smashwords, 2015, Book 3, ebook. CreateSpace, 2017, paperback)

Book 4. *Maritime Watchstanding Plans: Origins, Variants and Effectiveness* (Smashwords, 2015, ebook. CreateSpace, 2017 paperback)

Maritime Watchstanding Plans: Origins, Variants and Effectiveness

Preface

This book is the most comprehensive reference work available concerning (1) the genesis and history of maritime watchstanding and (2) more than a half-century of research concerning different watchstanding plans. The book includes assessments of more than 35 watchstanding plans that have been observed in civilian or military operations and/or studied in laboratories. Reference is made to 331 technical publications.

My interest in creating this reference work stemmed from my involvement in several research investigations of fatigue and performance in the maritime environment across portions of four decades (100, 189, 207, 213, 239, 244, 246, 248). I have also written about shiftwork scheduling and about fatigue as a contributor to the occurrences of accidents and incidents (196, 197).

The book is divided into three sections. The first section summarizes available information about maritime watchstanding practices from ancient times through the 1800s. This historical summary includes relevant information about the development of the measurement of time, especially at sea. The second section provides reviews of watchstanding research literature and summarizes the objectives, methods, results, and lessons learned from my own and others' investigations of the effects of watchstanding plans on mental performance. In the third section, I have summarized about 25 years of recommendations for fatigue risk management systems (FRMS) for maritime operations, and then presented some concluding thoughts.

Structure

Section I of the book focuses on the history of the development of watchstanding, especially at night. Chapter 1 addresses "The Night Watch in Ancient Times," with discussions of the concepts of kairological and chronological time and of informal and nautical astronomy Chapter 2 focuses on the development of the use of the clock at sea and of watchstanding at sea. Chapter 3 describes what may be called "classic" watchstanding plans. The chapter includes descriptions of dogging the watch, the use of the ship's bell to signal the time and the change of watch, the naming of the "first" watch, the practice of standing during the watch, and staying alert during the watch.

Section II focuses on research concerning watchstanding. It starts with primers on sleep and circadian rhythms. Then progresses to an annotated and commented bibliography separated into:

> Surface Watchstanding Studies 1950-1969 (Chapter 4)
>
> Surface Watchstanding Studies 1970-1989 (Chapter 5)
>
> Surface Watchstanding Studies 1990-1999 (Chapter 6)
>
> Surface Watchstanding Studies 2000-2015 (Chapter 7)
>
> Research Belowdecks (Chapter 8) with discussions of the uses of bright light therapy and protective eyeglasses
>
> Submarine Watchstanding Studies 1947-1999 (Chapter 9)
>
> Submarine Watchstanding Studies 2000-2015 (Chapter 10)

Section III contains an annotated and commented bibliography on modern maritime fatigue risk management systems (FRMS; Chapter 11) and then my concluding thoughts on wachstanding research. This is followed a by a numbered list, alphabetical by author, of the 331 technical references cited in the book. Numbers in parentheses at the ends of sentences refer to works cited in the reference section. The ToC indication after a subsection heading is a hyperlink back to the Table of Contents.

I prepared an appendix titled 'Watch Plan Analyses and Detailed Analysis Results.' The appendix describes the methods I used to analyze the nominal watch plans shown in this book. I used the Sleep, Activity, Fatigue, and Task Effectiveness (SAFTE) model and its software implementation, The Fatigue Avoidance Scheduling Tool (FAST) to model the effects of each watchstanding plan shown in this book quantitatively (48, 136–141, 198). Of these various metrics, I reported in the text of the book those that might be of interest to the reader. A complete set of metrics is available in the full Appendix. The Appendix is available from the author, and I have also posted it on ResearchGate.

Maritime Watchstanding Plans: Origins, Variants and Effectiveness

Table of Contents

Preface
SECTION I. HISTORY
Chapter 1. The Night Watch in Ancient Times
  Kairological Time
  Chronological Time
  Informal and Nautical Astronomy
  The Night Watch
Chapter 2. The Clock at Sea
  Hours of Equal Length
  Watchstanding at Sea
Chapter 3. "Classic" Watch Plans
  Watch Plan Structure
  The Classic Plans
  Dogging the Watch
  Classic Dogged Watch Plans
  Ships Bells
  Naming of the "First" Watch
  Standing during the Watch
  Keeping Watch during the Watch
SECTION II. RESEARCH
  Sleep Primer
  Circadian Rhythm Primer
Chapter 4. Surface Watchstanding Studies 1950-1969
Chapter 5. Surface Watchstanding Studies 1970-1989
Chapter 6. Surface Watchstanding Studies 1990-1999
Chapter 7. Surface Watchstanding Studies 2000-2015
Chapter 8. Research Belowdecks
  The Use of Bright Light Therapy
  The Use of Protective Eyeglasses

Chapter 9. Submarine Watchstanding Studies 1947-1999
Chapter 10. Submarine Watchstanding Studies 2000-2015
SECTION III. FATIGUE MITIGATION
Chapter 11. Modern Maritime FRMS
Chapter 12. Concluding Thoughts on Watchstanding Research
References
Appendix. Watch Plan Analyses and Detailed Analysis Results
About the Author

List of Figures

Figure 1. Possible Tower of Babel, a representation of astronomic knowledge.
Figure 2. Gerbert d'Aurillac (Pope Sylvester II) and the astrolabe attributed to Gerbert.
Figure 3. Prince Henry the Navigator.
Figure 4. The mariner's astrolabe.
Figure 5. Nominal alignment of the three-and four-watch systems of the 1st century AD.
Figure 6. Base-12 counting on one hand.
Figure 7. The Liturgy of the Hours.
Figure 8. Balance wheel in a cheap 1950s alarm clock.
Figure 9. John Harrison.
Figure 10. Ship's bells, per The British Horological Institute.
Figure 11. Rough approximation of an eight-hour sleep histogram.
Figure 12. Relative real-job speed and accuracy measures across the hours of the day.
Figure 13. Percent of shipping collisions at each hour of the day.
Figure 14. Matt Smith (left) and the author at Cold Bay, Aleutian Islands.
Figure 15. Sleep loss, sleeping, fatigue and accidents.
Figure 16. Circadian pattern of 23 groundings.
Figure 17. Numbers of hours slept per day as a function of rank and as a function of department.
Figure 18. Previous watchstanding experience.
Figure 19. Total sleep per 24 hours observations made by master's students at NPS.

Figure 20. Comical tree planting "ceremony" at the 430th AAFRTU, Ephrata Army Air Base, 1944.

Figure 21. Groups used for comparison in the 1949 submarine study.

Figure 22. Preferences about the close 4 plan.

Figure 23. Submariners' circadian cycle lengths.

Figure 24. Responses of 143 enlisted submariners about their experience with various watch plans.

Figure 25. NSMRL fixed 8 research plan.

Figure 26. Mean and minimum cognitive effectiveness over days at sea for both of the 1-and-2 back watch periods.

Figure 27. Mean and minimum cognitive effectiveness for each of the seven 1-and-3 watch periods.

Figure 28. Example of results produced by a CEWG during their analysis of a commercial maritime environment.

Figure 29. Comparison of watch plans.

Figure 30. Three-factor alertness model from the Horizon Toolkit report.

Figure 31. Screen shot of MARTHA interface.

SECTION I. HISTORY
Chapter 1. The Night Watch in Ancient Times

To establish and maintain a rotating watchstanding system, you need a way to tell time. In these first two chapters, I address the parallel developments of watchstanding and the measurement of chronological time. Being neither historian nor horologist, I have probably left out some useful details and references about the development of time keeping (horologist: Latin *horologium* from Greek ὡρολόγιον, from ὥρα, *hōra*, "hour or time" plus -o- plus suffix *logy*; literally, the study of time). I have focused on Middle Eastern and Western developments in time keeping. I have used the term "watchstanding" throughout this book; I view the term "watchkeeping" as its equivalent.

Kairological Time

There are two ways to tell time. The following characterization of the ancient Greek approach is somewhat generalizable to other societies.

> *Kairos* (καιρός) is an ancient Greek word meaning the right or opportune moment (the supreme moment). The ancient Greeks had two words for time, *chronos* and *kairos*. While the former refers to chronological or sequential time, the latter signifies a time lapse, a moment of indeterminate time in which everything happens. What is happening when referring to *kairos* depends on who is using the word. While *chronos* is quantitative, *kairos* has a qualitative, permanent nature. *Kairos* also means weather in both ancient and modern Greek. The plural, καιροί (*kairoi* (Ancient Modern Greek)) means 'the times.' (*Wikipedia*, March 2014; agrees with other sources)

The kairological method of describing events was used extensively in the Old Testament and in the Gospels of the New Testament. Thus, we know that this method has been in use for the last 4000 years or so. Randolph Richards and Brandon O'Brien reported,

> The ancients used *kairos* to refer to the more qualitative aspect of time, when something special happened. This term is used much

> more often-almost twice as frequently-in the Bible. Sometimes translated "season," *kairos* time is when something important happens at just the right time. (267)

> In the non-Western world, by contrast, the correct time is often connected to a condition or situation. Some call this an "event" orientation, in which, as Duane Elmer writes, "Each event is as long or as short as it needs to be. One cannot determine the required time in advance. Time is elastic, dictated only by the natural unfolding of the event. The quality of the event is the primary issue, not the quantity of minutes or hours." Relationships trump schedules, so things begin when everyone who needs to be there has arrived. (267)

The use of kairological time continues to this day.

> It took me years to realize that *siang* was connected to the temperature, not the clock. Once the morning had turned hot, it was *siang*. When it cooled down in the afternoon, it was *sore*. (267)

I present some research reports about watch plans later in this book kairologically; that is, out of chronological sequence but placed to help the reader make a logical connection. Additionally, I present research concerning watchstanding on surface ships separately from research concerning watchstanding on submarines, even though they occurred concurrently. Again, this was to make logical connections within these two differing work environments.

Working 24/7. Shore-based operations that occurred twenty-four hours per day, seven days per week (24/7) date back many thousands of years to sheepherding by middle-eastern nomadic tribes, such as the Bedouin. There was only one "watch section" for this 24/7 operation: the sheepherder who lived full-time with the flock; a practice that continues to this day in Wyoming, where I live. Martin Nilsson of the University of Lund suggested that in ancient times, days and nights tended to be viewed as separate entities, and ancient societies probably did not have words that encompassed the pairing of both "day" and "night" as a unit (236).

However, we do learn through Moses in *Genesis* 1:5b, "And there was evening and there was morning, the first day" (*English Standard Version, or ESV*). Historical/contextual Bible scholars note that these were 24-hour days, not longer periods. This phrase is reiterated in verses 1:8b, the second day, 1:13, the third day, 1:19, the fourth day, 1:23, the fifth day, and 1:31b, the sixth day. Harrison Cowan, an historian with the Longines-Wittnauer Watch Company, noted that *Genesis* revealed "the custom of counting the passage of 'days' by nights." This custom of marking "days" was used later especially in Indo-Europe and the Americas; the passage of dawns was used less often (Nilsson, op. cit.). This would be a somewhat kairological approach to telling time.

The division of the day into periods by primitive societies was based somewhat kairologically upon natural phenomena (Nilsson, op. cit.). Nilsson provided many examples, including the herding of sheep and cattle. For example, the Banyankole of Uganda reportedly specified 6 am, milking time; 9 am, (not translatable); noon, rest for the cattle; 1 pm, time to draw water; 2 pm, time for the cattle to drink; 3 pm, the cattle leave the watering place to graze; 4 pm, the sun shows signs of setting; 5 pm, the cattle return home; 6 pm, the cattle enter the kraal; 7 pm, milking time.

The first example Nilsson provided of a primitive society dividing the night into periods somewhat kairologically came from the Nandi of east Africa: 7 pm, the heavens are fastened; 8 pm, the porridge is finished; 9 pm, those who have drunk milk are asleep; 10 pm, the houses have been closed; 11 pm, those who sleep early wake up; midnight, the middle of the night. However, in most primitive societies, the night was just divided into three components: sunset to midnight, around midnight, and midnight to cock-crow or dawn.

Nilsson reported. "On the Marquesa Islands the first night-watch was 'the hour of ghosts'; the advanced night was termed 'black night', and midnight 'great sleep'; the last watch of the night was 'the coming of the day'. The Wadschagga [of east Africa] have three night watches: the awakening in the evening, that in the middle (midnight), and that in the morning twilight." The Tahitians divided both day and night into six periods each.

Nilsson's expressed assumption was that, like the primitive societies of Nilsson's day, very ancient societies may have used similar methods for dividing the day and the night into periods. Actually, many ancient societies were much more sophisticated. They used astronomy to define time chronologically, though their estimates of the passing of the seasons were probably much more accurate than their estimates of periods of the night. Even so, it seems safe to say that the human ability to estimate chronological time with reference to the stars pre-dates written history and existed in parallel with kairological approaches.

Chronological Time

The sun, the moon and the stars provided a great deal of chronological information to ancient societies about the time of day and night. The practices of both formal and especially informal astronomy developed in many ancient cultures. Nilsson noted that the Homeric Greeks practiced informal astronomy, as did Laplanders, Eskimos, and primitive societies in the Polynesian Islands and South America (236).

The length of the stellar day is slightly shorter than the length of the solar day. Our physical view from Earth of the stationary stars precesses forward by 3 minutes 56 seconds per day with respect to our view of the sun. Thus, some of the stars appear to move toward the west. Ancient and some recent primitive civilizations were aware of this phenomenon of precession. They could ascertain the passage of the seasons by relating the relationships between the rising or setting of certain stars to the rising or setting of the sun. Regarding this phenomenon, Nilsson observed cleverly that, "The stars are so to speak the stationary ciphers on the clock-face and the sun is the hand" (236).

Notable in the night sky, the Zodiac is made up of twelve constellations of stars along the ecliptic. The ecliptic is the apparent path of the sun around the earth, offset by about 23 degrees from the equator. These twelve constellations are about 30 degrees apart on the ecliptic (twelve constellations times 30 degrees gives a 360-degree circle). Thus, a constellation of the Zodiac appears above the night horizon approximately every two hours.

Presumably, the Sumerians were using the Zodiac by 3000 BC. Additionally, they used a calendar with a seven-day week, divided the 24-hour day into twelve periods and divided each period into 30 parts (about four of our minutes each). According to Harrison Cowan this latter knowledge about time of day did not transfer to directly other cultures (79). Sumeria was the ancient, non-Semitic culture of Mesopotamia that developed before 4000 BC. Chaldea was the Semitic area in the south of Mesopotamia, where the Euphrates River empties into the Persian Gulf. The Semites are said to be the descendents of the Biblical family of Shem, son of Noah, while the Sumerians may have been descendants of Nimrod, great-grandson of Noah through Noah's son, Ham, and Ham's son, Cush. The Chaldeans integrated with the Sumerians and became dominant in Mesopotamia as the Akkadian Empire.

Anglican clergyman Ethelbert Bullinger published a landmark book in 1920, *The Witness of the Stars*, in which he compiled a large amount of information about ancient astronomy. (Thank you to Pastor Duane Simonson for bringing this book to my attention.) Bullinger worked in part from his own research, from ancient writings and from previous research presented in Frances Rolleston's *Mazzaroth* (1865) and *Mizraim; Or, Astronomy of Egypt*, Joseph Seiss' *The Gospel in the Stars* (1884), and others (35). Bullinger credited Rolleston with the acquisition of data from "Albumazer, the Arab astronomer to the Caliphs of Grenada, 850 A.D.; and the Tables drawn up by Ulugh Beigh [Mīrzā Muhammad Tāraghay bin Shāhrukh, or Ulugh Beg], the Tartar prince and astronomer, about [AD 1437], who gives the Arabian Astronomy as it had come down from the earliest times" (pp. iii-iv).

The basic point made by Bullinger and some predecessors was that the constellations of the Zodiac were placed and named by God: "The heavens declare the glory of God, and the sky above proclaims his handiwork" *(Psalm 19:1, ESV)*. Bullinger deconstructed *Psalm 19* to show its paired astronomical and literary references to ancient astronomy. He also extracted astronomical references from *Job*, which was then and is still suspected to be the oldest book in the Bible (104, pp. 781-782). The book of *Job* may pre-date 2268 BC, the approximate time of the Flood by my count of the life spans of the generations specified in the Bible (ibid., p. 808, note 20:24). The Hebrew in *Job 9:9* referred to the constellations *Ash* (Arcturus), *Cesil* (Orion), and *Cimah* (Pleiades). *Job 38:32* referred to the *Mazzaroth*, all of the constellations.

Bullinger noted that the word *Zodiac* has no relation to the animal names used for the constellations of the Zodiac in modern times. Instead, it derives from "a primitive root through the Hebrew *Sodi*, which in Sanscrit means *a way*. Its etymology ... denotes *a way*, or *step*, and is used of the *way* or *path* in which the sun appears to move amongst the stars in the course of a year" (ibid., p. 15).

Bullinger also noted that Chinese, Chaldean and Egyptian records pre-dating 2000 BC show essentially identical Zodiacs. He suggested that many of these presentations of the Zodiac indicate knowledge of it around 4000 BC, when the summer solstice was in Leo. This dating is allowed by back-calculation based upon the precession of the equinoxes at a rate of about one degree per 71.6 years, and would place the knowledge of the Zodiacal constellations within the Sumerian culture.

According to Bullinger,

> While *Alpha* in the constellation of *Draco* was the Polar Star when the Zodiac was first formed, the Polar Star is now *Alpha* in what is called *Ursa Minor*. This change alone carries us back at least 5,000 years [about 3000 BC]. The same movement which has changed the relative position of these two stars has also caused the constellation of the *Southern Cross* to become invisible in northern latitudes. When the constellations were formed the *Southern Cross* was visible in N. latitude 40 degrees, and was included in their number. But, though known by tradition, in had not been seen in that latitude for some twenty centuries, until voyages to the Cape of Good Hope were made. (ibid., pp. 14-15)

Rolleston in *Mizraim* showed the ancient Coptic names of the twelve signs of the Zodiac, according to Ulugh Beigh and then noted:

> For the first 2000 years of the Hebrew chronology the summer solstice took place in Leo. After perhaps about 1700 years of that time, Egypt was settled and civilized, preserving prophetic and astronomical traditions from the Antediluvians, through Noah and Ham [Sumeria?], their more immediate ancestors, to which these names testify. In the first thousand years of that time the inundation of the Nile occurred, while the sun was still in Leo, at the summer solstice; to this time then the origin of these names must be referred, where *Pi Mentekeon*, the pouring out, is translated *Cubitus Nili*.

Bullinger asserted, "Ancient Persian and Arabian traditions ascribe [astronomy's] invention to Adam, Seth [Noah's ancestor] and Enoch" (ibid., p. 10). He also presented an archeological case for the top of the Tower of Babel having been a representation of the astronomic knowledge of heavens, i.e., sun, moon and stars, constructed as an instrument for preserving this information beyond the Flood. Though known to archeologists since the mid-1800s, the potential Babel tower, *Birs Nimrud* (ancient: *Borsippa*) was excavated only partly in the 1980s and was looted in the post-2001 Iraq war (Figure 1). It is located eleven miles southwest of Babylon, on the east bank of the Euphrates, and is at least as old as the Ur III period of 2112-2004 BC, not that far off from the apparent date of the Flood. Looted tablets have turned up supporting the idea that this tower held astronomical information.

Figure 1. Possible Tower of Babel, a representation of astronomic knowledge. (Wikipedia)

Bullinger noted that the first century AD Jewish historian Josephus, working for the Romans, mentioned the ancient tradition of astronomy among the Jews. In modern translation, Josephus stated, "The virtuous descendants of Seth discovered the science of astronomy" (147). Additionally, according to Josephus, "Berosus states: 'In the tenth generation after the flood, a great and just man lived among the Chaldeans, well versed in celestial lore' " (ibid.). This was Abram (1976-1801 BC), renamed Abraham by God and from whom the Arabs, Jews and Jesus Christ were descended. Apparently, Abram was an influential man in both Ur, the main city of Chaldea, and then in Damascus. (According to Wikipedia, Berosus was a Chaldean priest of Bel (Baal) in Babylon who was most active around 290 BC, and who wrote about the history and culture of Babylonia. Abram spent time in Babylon and Damascus while enroute from Ur in Chaldea to Canaan, as described in *Genesis*.)

By about 2100 BC, Sargon created and was the ruler of the Semitic Akkadian Empire, which encompassed Ur in the south and the Levant north of Ur, as far as Ninevah. Seventy astronomical tablets, now in the British Museum, were created at the direction of Sargon. They are known as *The Illumination of Bel* or *When the Gods Anu, Bel*. Thus, we know that astronomy was a strong science at the time of Abram. In fact, Abram's views of astronomy may have been one basis for his subsequent emigration from Chaldea to Canaan. Josephus stated, "He [Abram] was the first to declare boldly that God [not Bel], the creator of the universe, is one, and that the sun, moon and stars had no inherent power of their own. Because of these opinions the Chaldeans rose against him, and so he emigrated to Canaan with God's help" (ibid., p.24).

Subsequently, the Kassite dynasty of Babylon, up the Euphrates River from Ur, began around 1650 BC. Nilsson asserted that the use of a "double hour" arose in the Kassite culture (236). The double hour was based upon the approximately-two-hour interval between appearances above the horizon of the twelve constellations of the Zodiac, six of which are, by their careful selection, always present in the night sky. Obviously, the informal astronomer could use the double hour to divide the nominal twelve hours of nighttime into three four-hour watch periods. The Kassites were militant, so it is likely that they would have had an interest in the careful scheduling of a night watch.

Though chronological specifications of periods within the night were made possible by formal and informal astronomy and by daily tasks, they had no reliable lengths. The period lengths varied widely. This variability probably (1) was inconvenient in daily life except perhaps near the equator, (2) made the development of the clock difficult, and (3) was impractical for the development of formal (scientific) astronomy (236). The ability and desire to enumerate the hours of the day night chronologically developed slowly across millennia, and then more rapidly across recent centuries.

Informal and Nautical Astronomy

Harrison Cowan noted,

> For countless years, time and timekeeping were the special domain of priests and royal people. Observation of the stars served the needs of the priest-astrologer. A general knowledge of timekeeping among ordinary people hardly goes back two hundred years. Such knowledge was unnecessary until the social organization of the world began to approach the tightly integrated society of today. (79, p. 15)

References that I have collected and cited in this book suggest that the two-hundred year number probably did not apply to those associated with the night watch nor to seafarers. The knowledge of timekeeping dates to much earlier periods for watchstanders, but these were not "ordinary" people.

Nilsson used Homer to characterize the Greek approach to telling the time of day and night (236). The Homeric literature was probably transcribed at some point in the period 600 to 800 BC from older oral traditions, and included knowledge gained from Babylonia. Achilles speaks of dawn (forenoon), noon and afternoon. Also mentioned are the twilight periods, the appearance of the morning and evening star (*Venus*), and the division of the night into three parts with reference to the stars (informal astronomy). Though never specified in Homer, those three parts were probably four-hour watches.

Moving forward in time and geographically to the west, the spread of informal astronomy is illustrated by the popularity in Greek and then Roman cultures of two poems by Aratus, *Phainomena* and *Diosemeia*, each of which addressed general knowledge of the stars and were written just after 276 BC. (Of course, some of the popularity dealt with the then-millennia-old use of astrology to predict the future, but that subject is outside the bounds of the present book.) Bullinger made an interesting observation about the poems of Aratus: "Aratus describes them [the stars], not as they were seen in his day, but as they were seen some 4,000 years before. ... he must therefore have written from a then ancient Zodiac" (35, p. 14). In the first century AD, Saint Paul, previously a student of Gamaliel, was likely to have learned some aspects of informal astronomy. According to *Luke*, Paul quoted the fifth line of Aratus' *Phaenomena* in his first century address on the *Areopagus* in Athens when he spoke of the unknown god of the Greeks (*Acts 17:28*): "For we are indeed his offspring" (ESV).

Through the centuries, various ancient devices were invented that allowed measurements of time of day and of latitude. The water clock had appeared before 1000 BC, predating the Greek and Roman periods. It is possible, but not proven, that the Greeks and Romans may have used hourglasses; these devices bacame common many centuries later. The Greeks, perhaps Hipparchus, created an early astrolabe in about 150 BC. An astrolabe is a "star-taker," the name being derived from Medieval Latin to the Greek word *astrolabos*, from *astron* "star" and *lambanein* "to take." In Arabic texts the word *asturlab* is translated as *akhdh al-kawakib*, or "taking the stars."

A geared astronomical instrument was recovered in 1900-1901 from the Antikythera wreck, a shipwreck off the Greek island of Antikythera. Inspection revealed that the instrument was designed and constructed by Greek scientists in the approximate period 150 to 100 BC. The device predicted astronomical positions, eclipses and the cycles of Olympic Games (110, 111). Around 250 BC, Archimedes had designed gear mechanisms for water clocks to generate mechanical motions, providing the general basis for all gear design (79). The twelve months labeled on the Antikythera mechanism appeared to be of Corinthian origin, "suggesting a heritage going back to Archimedes" (111).

One may read more about this device at the website of the Antikythera Mechanism Research Project (http://www.antikythera-mechanism.gr/, 15 December 2014). According to that web site, "Nothing as complex is known for the next thousand years." I wonder if this was a one of a kind device that married the recent development of gear design with ancient knowledge of astronomy and, having been lost at sea, could not be reconstructed for some unknown reason. (Thank you to friend and author, Steve Rzasa, for bringing this device to my attention.)

So, the Roman Empire (Caesar Augustus, 27 BC, to Romulus Augustus, AD 476) and the Roman republic that had preceded the Empire for 500 years applied Greek astronomy to timekeeping, astrology and mythology. With the end of the Empire in the fifth century AD, local bishops of the western (Roman) Christian church served as *de facto* governors, while invaders from the north occupied the western lands that had been controlled by the Empire (105, pp. 110-111). As far as I can tell, there were no significant advances in timekeeping or navigation sciences in Europe for 500 or 600 years after the fall of Rome.

The next significant developments in astronomy and timekeeping came from Muslim scientists. A century and a half after the end of the Roman Empire, the Prophet Mohammed died in AD 632. "Led by the earliest Muslim caliphs ... Muslim forces spread out explosively in all directions. They had conquered Syria, Iraq and Jerusalem by 637, Egypt by 642, Central Asia and western north Africa by 670. Less than fifty years later the armies of Islam had invaded Spain, Persia, and India and were conducting raids across the Pyrenees" (306).

Howard Turner, science curator for the 1982-83 exhibition, *The Heritage of Islam*, described the developments in astronomical science that occurred subsequently in medieval Islam (306). Islamic scientists studied the apparent movements of the heavenly bodies. They extended the astronomy developed by Greek, Ptolemaic, Indian, and Sasanian philosophers. Other than Apollonius, virtually all of the ancient astronomers had placed the earth at the center of the planetary system. While medieval Islamic scholars continued to share this view, "Slowly but surely the long-established canons of classical astronomy were being challenged by Muslim attempts to force Aristotelian and Ptolemaic theories into a practical, functioning system that described what really went on in the space surrounding the earth" (306). Ibn al-Shatir was a leading innovator in this respect.

Turner attributed the perfection of the astrolabe and its simpler cousin, the quadrant, to astronomers in medieval Islam, including the invention of the quadrant by Muslims in Egypt in the 11th or 12th century (306). He noted that by the 16th century the quadrant had replaced the astrolabe in the Muslim world except in Persia and India. Presumably, the quadrant was used to help identify the five daily prayer times in the "science of fixed moments," or *'il al-miqat*. In this branch of astronomy, inherited from the Babylonians and Egyptians, the Islamic day began at sunset and the Islamic month at the new moon. The quadrant was not as practical for use at sea as the astrolabe, as discussed below.

Citing sources including the UNESCO book, *Science and Technology in Islam* (127) and the review by Turner (306), writer Zakaria Virk of Toronto, Canada, noted:

> The astrolabe was an astronomical instrument made of brass. It served as a sort of computer for nearly a thousand years. It provided data for calculating latitude, determining time, surveying, nautical observation, and the casting of horoscopes. Muslim interest in this instrument was immense due to the fact that they could determine the hour of canonical prayer, direction of Mecca, and beginning of new months. It was introduced in Europe in the 10th century by Gerbert (later Pope Sylvester II from AD 999 to 1003) who learned its use during his sojourn in [Islamic] Cordoba. It was used in Europe until 16th century for nautical observation. Every astronomer whether Jewish, Christian, or Muslim used it. "How to Use" manuals were written in Arabic, Hebrew, and

Latin. The earliest such manual was written by Mashallah in Iraq around AD 800. (313, 314)

Gerbert may have been the author of a description of the astrolabe that was edited by Hermannus Contractus some 50 years later (Figure 2). I've written more about Gerbert, below.

Figure 2. Gerbert d'Aurillac (Pope Sylvester II) and the astrolabe attributed to Gerbert, ca. AD 1000 (Wikipedia). Apparently based upon Islamic designs.

In al-Andalus, Hasday ibn Shaprut obtained Mashallah's treatise on Astrolabe (*Ar. Asturlab*) that became the basis of all future treatises. Maslama al-Majriti wrote a treatise that was translated by John of Seville in 12th century. There is an astrolabe at Oxford Museum dated 984 made by Ahmad ibn Ibrahim of Isphahan, Iran. The astrolabe used by Chaucer can be seen at Merton

College Library, Oxford. One made by Ibrahim ibn Saeed at [Muslim] Toledo in 1066 can be seen at Victoria & Albert Museum, London. One made by [a] Muslim instrument maker of Valencia is at [the] Smithsonian Institution... The description of [the] as-Safiha astrolabe made by Al-Zarqali was widely read in 13th century in Castilian. Portuguese explorer Vasco da Gama's ships [ca. 1500] were fitted out with astrolabes designed by a Spanish astronomer. (313, 314)

The next contirbutions to navigation technology were made by the Potruguese. The mariner's astrolabe was well known by the end of the 15th century, with creation and perfection by Portuguese navigators during the beginning of the Portuguese seafaring discoveries of the 15th and 16th centuries AD. Astronomy had been taught in Portugal for a long time:

> *Libros del Saber Astronomia* (1252), an important work on astronomy written by great peninsular astronomers, co-ordinated all the knowledge of astronomical science till then. Besides many astronomical details, it described the nautical instruments, their construction, and dealt with various aspects of navigation by astronomy. This work had exercised considerable influence on the Portuguese nautical science. The University of Coimbra, founded in 1290 by King Diniz, also encouraged the study of astronomy in Portugal. (186)

Portuguese nautical science began mainly with voyages by Prince Henry the Navigator (1394-1460; Figure 3) into the Atlantic and down the coast of Africa (298). The important roles to be played by the Portuguese were to combine the astronomical and nautical sciences and then make the results of the collaboration available (186). (Thank you to André Pimenta for researching the Portuguese navigation references.)

Figure 3. Prince Henry the Navigator.

Prince Henry established a school at Sagres, where navigational methods and instruments were developed. According to K.M. Mathew,

> The Portuguese mariners used two types of nautical instruments: (i) astrolabe and quadrant which gave them directly the angular height of the star observed, (ii) and balestilha [cross-staff] and tables of India which gave it by means of two linear elements. (186)

The quadrant was difficult to use on board a moving ship and did not provide all of the needed functions. The school then developed a mariner's astrolabe toward the end of the 15th century. According to Alan Neale Stimson,

> The earliest recorded use of the sea astrolabe was made during a voyage undertaken in 1481 by Diogo d'Azambuja down the West coast of Africa. Bartholomew Diaz on his voyage to the Cape of Good Hope in 1487-88 was equipped with astrolabes while Vasco de Gama is recorded as using an astrolabe three 'hands' in width (about 24 inches) suspended from a trestle, at St. Helens Bay, South Africa in 1497. This, and some small astrolabes had proved difficult to use at sea. In 1519 Magellan is also recorded as being equipped with one wooden and six metal astrolabes. (298)

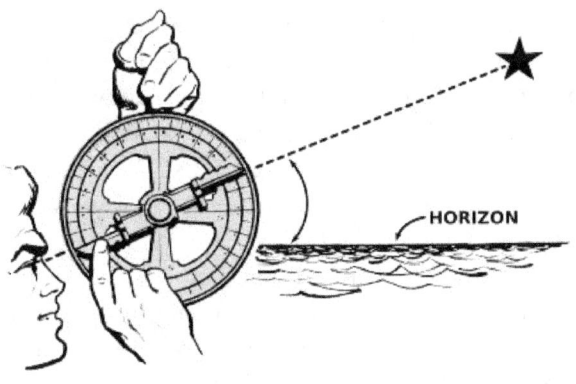

Figure 4. The mariner's astrolabe.

The mariner's astrolabe, or sea astrolabe or sea ring, was not an astrolabe proper (Figure 4). It was a graduated circle with an alidade (a "turning board" that allows one to sight a distant object and use the line of sight used to measure vertical angles).

According to Stimson,

> The sea astrolabe continued in use throughout the 17th century but increasingly northern European seamen began to favour the cross-staff and Davis quadrant or backstaff (invented about 1590 by Captain John Davis, the English Arctic explorer) for solar observations. ... In France the instrument was in use at least until 1690. (298)

The mariner's astrolabe was cumbersome because it had to be suspended and often produced large, unavoidable errors in rough weather (223). Some sources indicate that the development of the cross-staff (Jacob's staff, *ballastella*) supplanted the use of the astrolabe. Other sources indicate that the cross staff was used in Chaldea as early as 400 BC. Designed to measure angles, the cross-staff consisted of a long piece of wood with a crosspiece that could slide along a graduated staff. The navigator would sight a celestial body along the top of the crosspiece and move the cross piece so that the bottom tip met the horizon. The angle of the body above the horizon could be read on the staff. Navigators often used the sun as a sight, and the use of the cross-staff often caused eye damage for longtime navigators.

The back-staff was developed in about 1590 so that navigators didn't have to look into the sun. The navigator faced away from the sun and used the shadow cast by the device to measure the angle between the body and the horizon. Captain John Davis revised the backstaff in 1594. The Davis quadrant had a physical arc affixed to a staff such that it could slide along the staff. He placed the arc so that it cast its shadow on a horizon vane. The navigator would look along the staff and observe the horizon through a slit in the horizon vane. By sliding the arc so that the shadow aligned with the horizon, the angle of the sun could be read on the graduated staff. Davis improved the quadrant substantially by the mid-17th century. By using two arcs, he made it both accurate and portable. The Davis quadrant was one of the most widely used forms of the backstaff, and Continental European navigators called it the English Quadrant.

Later, smoked mirrors were added so that bodies which were too dim to give a shadow could be used while at the same time, dimming the reflection of the sun so that it could still be used. This foreshadowed the modern sextant. (223)

> The modern sextant was invented by Sir Isaac Newton in 1700. However, he never built one. It was an Englishman named John Hadley and an American named Thomas Godfrey who built two different versions of the sextant, both equally useful. Later, Paul Vernier added a second graduated arc to make measurements more precise. Even later, the micrometer screw were invented, which further added to the precision of the instrument and became what we now recognize as the modern sextant. (223)

I leave it to the reader to pursue knowledge about the modern day sextant. However, I digress briefly here with a personal anecdote. I attended the one-year USAF undergraduate pilot training (UPT) course in 1966-1967 at Webb AFB, Big Spring, Texas. The USAF had many pilot training bases operating in the U.S. during the Vietnam War, and each base graduated a new class of about 40 officers every six weeks. At a graduation a formal military parade was held, during which pilot wings were pinned on the new graduates' uniforms and we were awarded our aeronautical ratings. Traditionally, there would be a fly-by of a Northrop T-38 Talon trainer, which we had been flying, or another type of USAF aircraft, and there would also be a static display of various operational USAF aircraft. One reason for the static display was to allow student pilots to inspect the airplanes and to speak with operational aircrews about their missions. This helped us decide what airplane assignments to bid for upon graduation.

I was drawn to the Lockheed C-130E Hercules (Herk). (Subsequently, I flew 2,000 hours in the Herk during 1967-1971, including about 700 hours of combat time). During a UPT graduation early in 1967, I climbed into the cockpit of the static-display Herk and engaged the aircrew in conversation. (Fortuitously, they were members of the squadron to which I was later assigned, and they later became crewmates and friends.) During the conversation I notice a round port of some sort in the roof of the cockpit. "What's that?" I asked. "That's the sextant port." I was stunned. I associated sextants with sailing ships of the 18th and 19th century, but I was standing in the highly-instrumented cockpit of a modern warplane of the mid-20th century.

When we flew over the ocean in the C-130E, we had limited navigational aids (navaids). If our short-wave system (high-frequency, or HF) could receive two Long-Range Navigation (LORAN) stations, we could triangulate our position reasonably well. However, we could not always receive, especially if our HF antennas iced up. The back-up system was the sextant, which our navigator could use for sun or star shots if we were not in clouds.

The back up to these two methods was dead reckoning, which is also used extensively by seafarers. This means, basically, correcting for the wind (and ocean currents for seafarers) to proceed along a given course, and knowing how the wind and other factors affect one's ground speed. When we had been using dead reckoning for many hours and were about to penetrate an air defense zone near land, we would declare "minimum navaids" over the radio so that the radar installations looking for us to enter would know that we were using dead reckoning, might be a bit off course and not make our planned arrival time at the outer edge of the zone. Dead reckoning worked quite well for us because we had a look-down Doppler radar system that allowed us to make a very good estimate of the local wind, even when flying as high as 20,000 feet (6,100 meters). Nowadays, satellites provide excellent navigation coverage. However, they do not provide data about an aircraft's position and track over much of the ocean, as evidenced by the mysterious loss of Malaysian Air flight 370 on 8 August 2014.

The Night Watch

Apparently, our Anglo-American word "watch" derives from Middle English *wacchen* or *wacan*, to awake, and Old English *wæccan*, to watch, or keep watch. The use of "watch" is recorded before the 12th century AD. One definition of the word, "watch," that I have found is, "Any of the definite divisions of the night made by ancient peoples." Other older definitions of "watch" include:

- The state of being wakeful.

- One of the <u>indeterminate</u> intervals marking the passage of night, usually used in plural; for example, the silent watches of the night. (emphasis added)

- Lookout, watchman.

- The office or function of a sentinel or guard.

Presently, we also use the word "watch" to mean several things. Definitions include:

- The act of keeping awake to guard, protect, or attend.

- A wake that is held over a dead body. (In some operations, this may be the dead body of the sentry who was found asleep while on watch!)

- A state of alertness and continuous attention.

- Close observation, surveillance.

- A notice or bulletin that alerts the public to the possibility of severe weather conditions occurring in the near future; for example, a winter storm watch.

- A portable timepiece designed to be worn on the wrist or carried in the pocket. The first watches appeared shortly after 1500, when the mainspring was invented as a replacement for weights to drive clocks.

- A holder, especially of an overseeing or managerial office; for example, the business grew on her watch.

Military and maritime definitions include:

- A body of soldiers or sentinels making up a guard.

- A period of duty.

- A portion of time during which a part of a ship's company is on duty.

- The part of a ship's company required to be on duty during a particular watch.

- A sailor's assigned duty period.

The Online Etymology Dictionary noted that the meaning, "period of time in which a division of a ship's crew remains on deck," is from AD 1585. The same source also reported that the sense of a "period into which a night was divided in ancient times" translates from the Latin *vigilia*, Greek *phylake*, and Hebrew *ashmoreth*.

Leaping forward to the present day, Title 46 of the U.S. *Code of Federal Regulations* states the following:

> Paragraph 15.705 Watches.
>
> (a) Title 46 U.S.C. 8104 applies to the establishment of watches aboard certain U.S. vessels. The establishment of adequate watches is the responsibility of the vessel's master. The Coast Guard interprets the term "watch" to be the direct performance of vessel operations, whether deck or engine, where such operations would routinely be controlled and performed in a scheduled and fixed rotation. The performance of maintenance or work necessary to the vessel's safe operation on a daily basis does not in itself constitute the establishment of a watch. The minimum safe manning levels specified in a vessel's COI or other safe manning document take into consideration routine maintenance requirements and ability of the crew to perform all operational evolutions, including emergencies, as well as those functions which may be assigned to persons in watches.
>
> (b) Subject to exceptions, 46 U.S.C. 8104 requires that when a master of a seagoing vessel of more than 100 GRT establishes watches for the officers, sailors, coal passers, firemen, oilers, and watertenders, "the personnel shall be divided, when at sea, into at least three watches and shall be kept on duty successively to perform ordinary work incidental to the operation and management of the vessel." Solely for the purposes of this part, the Coast Guard interprets "sailors" to mean those members of the deck department other than officers, whose duties involve the mechanics of conducting the ship on its voyage, such as

helmsman (wheelsman), lookout, etc., and which are necessary to the maintenance of a continuous watch. The term "sailors" is not interpreted to include able seamen and ordinary seamen not performing these duties.

Paragraph 15.710   Working hours

In addition to prescribing watch requirements, 46 U.S.C. 8104 sets limitations on the working hours of credentialed officers and crew members, prescribes certain rest periods, and prohibits unnecessary work on Sundays and certain holidays when the vessel is in a safe harbor. It is the responsibility of the master or person in charge to ensure that these limitations are met. However, under 46 U.S.C. 8104(f), the master or other credentialed officer can require any part of the crew to work when, in his or her judgment, they are needed for:

(a) Maneuvering, shifting berth, mooring, unmooring;

(b) Performing work necessary for the safety of the vessel, or the vessel's passengers, crew, or cargo;

(c) Saving of life onboard another vessel in jeopardy; or,

(d) Performing fire, lifeboat, or other drills in port or at sea.

(Title 46 *CFR*, Chapter I, Subchapter B, Part 15, Subpart G)

Now that we think we know what the "watch" is and how we derived the word, how did the night watch develop in ancient times? We know that from at least the time of the Exodus in the 1400s BC, the Jews had divided the night into watches:

> And in the morning watch the Lord in the pillar of fire and of cloud looked down on the Egyptian forces and threw the Egyptian forces into a panic (*Exodus 14:24, ESV*) (104)

We may infer from the occurrence of the Exodus that this night-watch practice may have been in use by the Egyptians, also, before 1400 BC. Subsequently, the first reference to "watchmen" in the Old Testament occurred in about 970 BC in the time of King Saul and in the areas of Philistine and Judah:

> And the watchmen of Saul in Gibeah of Benjamin looked, and behold, the multitude was dispersing here and there. (*1 Samuel 14:16, ESV*)
>
> Now David was sitting between the two gates, and the watchman went up to the roof of the gate by the wall, and when he lifted up his eyes and looked, he saw a man running alone. (*2 Samuel 18:24, ESV*)

King Solomon also referred to the night watch before about 931 BC.

> Unless the Lord builds the house, those who build it labor in vain. Unless the Lord watches over the city, the watchman stays awake in vain. (*Psalm 127:1, ESV*)

By about 587 BC the three watches were called the first or "beginning of the watches" (*Lamentations 2:19*), the middle watch (*Judges 7:19*), and the morning watch (*Exodus 14:24, 1 Samuel 11:11*). The latter was also called "Roostercrow." These occurred approximately from sunset to 2200, from 2200 to 0200, and from 0200 to sunrise.

It is possible that the people of Judah improved upon their scheduling of the nighttime three-watch system during their Babylonian exile in the 6th century BC. This was a century following a period in which the Neo-Babylonian culture excelled in developing astronomical science. Of course, the water clock (discussed below) was also available for the timing of the night watch after about 1000 BC.

Augustus Caesar helped found the Roman Empire and was its first emperor. He ruled that Empire from 27 BC until his death in AD 14. At that time, the *Triumviri Nocturni* kept watch over the city of Rome at night. Their work was reported as early as 304 BC (12). Their main duty was to walk around the city at night (*vigilias circumire*) to detect and fight fires. This privately operated system became ineffective, so in AD 6 Augustus created a public firefighting force called the *Vigiles Urbani* (153, pp. 154, 189).

Augustus levied a 4% tax ("There's nothing new under the sun." *Ecclesiastes 1:9, ESV*) on the sale of slaves and used the proceeds to set up the new force. He modeled this force after the fire brigade of Alexandria, Egypt (12). As noted above, the latter may already have had a 1400-year history. The *Praefectus Vigilum* was the chief of the night watch. His jurisdiction included some offences affecting the public peace, but he could apply only slight punishments. It appears that the *Vigiles Urbani* (or *Cohortes Vigilum*) used a three-hour, four-watch system at night: approximately sundown to 2100, 2100 to midnight, midnight to 0300, and 0300 to dawn (12).

In the first century AD, two night watch systems were cited in Judea: the Jewish three-watch system and the Roman four-watch system. However, the Jews may have cited the Roman system in some texts. For example, "And in the fourth watch of the night he came to them, walking on the sea" (*Matthew 14:25, ESV*). Other translations of this verse transliterate it to speak of a pre-dawn period instead of the fourth watch. However, according to my good friend, Pastor Gregory Truwe, the text indeed refers to a fourth guard or watch:

> It translates as: "But at the fourth guard/watch of the night he came to them while walking on the sea" (wooden translation). The word for guard/watch is *fulakh*, which in a different context is simply translated as prison (*Matt 5:25; 14:3, 10; 18:30; 25:36, 39, 43, 44*). In *Matt 24:43*, the same noun is translated as "part of the night" (*ESV*), or "time of night" (*NIV*). But in the specific text about which you have asked, the noun is modified by the adjective meaning "fourth," *te,tartoj*. By the way, the parallel passage in *Mark 6:48* also contains the same noun, modified by the same adjective, "fourth watch." In *Luke 2:8*, it was the shepherds who were guarding/watching the flocks, but of course, there is no time stamp there, just another example of guard/watch/prison having the sense of watch. *Luke 12:38*, however, uses the term "watch/guard/prison with both "second" and "third" modifiers. (Pastor Gregory Truwe, personal communication, 2014)

The association between these two watch plans in New Testament references was addressed by Troy Martin (184) and by Patrick Madden (180). Nominally, the two systems were aligned as shown in Figure 5.

Hour	Three-Watch	Four-Watch	Hour
1800 to 1900	Evening	First	1800 to 1900
1900 to 2000			1900 to 2000
2000 to 2100			2000 to 2100
2100 to 2200	Middle	Second	2100 to 2200
2200 to 2300			2200 to 2300
2300 to Midnight			2300 to Midnight
Midnight to 0100		Third	Midnight to 0100
0100 to 0200			0100 to 0200
0200 to 0300	Morning (Roostercrow)		0200 to 0300
0300 to 0400		Fourth	0300 to 0400
0400 to 0500			0400 to 0500
0500 to 0600			0500 to 0600

Figure 5. Nominal alignment of the three-and four-watch systems of the 1st century AD, used by the Jews and the Romans, respectively. (180, 184)

Harrison Cowan also suggested that "A watch was three or four hours long, depending on the time of year" (79). He cited no source for this information, but it seems logical that shorter watch lengths might have been used in the summer, when nights are shorter, or vice versa. The seasonal differences in the length of the night are greater, of course, as latitude increases. Jerusalem lies at about 32 degrees north. Thus, the length of the summer night there is about ten hours and of the winter night about 14 hours. You can do the watch-length math.

Sara Phang's research indicated that The Roman Army's "best-known chronological regimentation was the keeping of the night watches (*vigiliae*) in camp. The watches were divided into four equal parts of the night by the water clock, and *tesserae* (wooden tablets) with the watchword were circulated to ensure that all guards remained at their posts" (252). The circulating of tablets to night guards for this purpose was apparently still in practice in the late 2nd century AD.

With regard to the effects of fatigue on decision making, it is worrisome that in today's military there is a lack of concern about the sleep needs of a military commander. This lack may date back to the time of the Roman Army, with special reference to Socrates when he was an infantryman (288). As an example, Marcus Cornelius Fronto wrote around AD 163 about the effort Lucius expended in restoring order in the Roman Army.

"This great decay in military discipline Lucius took in hand as the case demanded, setting up his own energy in the service as a pattern. ... [H]e keeps the first watch easily, [and] for the last he is awake long beforehand and waiting" (112). So he might have slept only between about 2100 and 0300 or even much less.

How does the idea of standing a nighttime watch interact with the need for sleep? The National Sleep Foundation noted that research investigations of the last 80 years have led us to the conclusion that the best nighttime sleep is seven to nine hours long and uninterrupted. This sleep period maximizes mental performance during the daylight hours. However, what about a long winter's night that is 14 or 16 hours long?

Susan Brink cited archaic descriptions of sleep (27). She noted that as far back as Virgil and Homer, sleep was described as being divided into two periods during the night, separated by a period of wakefulness "sometimes called 'the watch'." Historian A. Roger Ekirch of Virginia Tech extended Brink's observation to medieval times, citing numerous sources for a normal period of wakefulness in the middle of the night (102, 103). For example:

> Until the modern era, up to an hour or more of quiet wakefulness midway through the night interrupted the rest of most Western Europeans.... . Families rose from their beds to urinate, smoke tobacco, and even visit close neighbors. Remaining abed, many persons also made love, prayed, and, most important, reflected on the dreams that typically preceded waking from their "first sleep."

Thus, given a long enough night, you are likely to divide sleep into two long episodes. However, long, dark nights are the exception in the modern era, especially in the presence of electric lighting sources. Long, dark nights are also an exception in today's multi-crew quarters on vessels that are underway. Crew quarters have lights on to accommodate crewmembers going on and off duty or relaxing during non-duty, non-sleep periods. This problem happens also in a cabin occupied by only two younger officers who are assigned to different watch sections. Also, with today's emphasis on minimal crew size, the sleep of off-duty crewmembers is often interrupted by unexpected problems that arise during ship's operations.

Interestingly, the night watch for cattle and sheep drives in Wyoming around 1915 was divided into four two-hour watches (316). The cowboys worked all day from 0400 until 2000. The first night guard rode out at about 2000. The second night guard relieved them at about 2200, the third night guard came on duty at midnight, and the fourth night guard worked from 0200 until 0400. A midnight supper was often available. It seems that two- to four-hour night watches have been popular selections for nighttime watchstanding duties.

Chapter 2. The Clock at Sea

Hours of Equal Length

Water clocks and sundials are among the oldest time-measuring instruments. The ancient Egyptians divided the day into two twelve-hour periods, and used large obelisks as sundials. The obelisks represented the sun god, Ra, and their presence dates back to 3100 BC (79). The shadows of the obelisks also allowed the marking of the summer and winter solstices and vernal and autumnal equinoxes (equal lengths of day and night). The Aztecs also developed this capability, as did other societies.

The obelisk was probably the basis for the story of the "Dial of Ahaz" in *2 Kings 20:8-11* in the Old Testament, dated to about 700 BC in Jerusalem:

> And [King] Hezekiah [son of King Ahaz] said to Isaiah, "What shall be the sign that the Lord will heal me, and that I shall go up to the house of the Lord on the third day?" And Isaiah said, "This shall be the sign to you from the Lord, that the Lord will do the thing that he has promised: shall the shadow go forward ten steps, or go back ten steps?" And Hezekiah answered, "It is an easy thing for the shadow to lengthen ten steps. Rather let the shadow go back ten steps." And Isaiah the prophet called to the Lord, and he brought the shadow back ten steps, by which it had gone down on the steps of Ahaz. *(ESV)*

This miracle may have been associated with a solar eclipse that occurred in 689 BC and was observable in Jerusalem. The sundial, onto which an obelisk would have cast a shadow, may date to the 6th century BC, but sundials with hours of equal length did not occur before about AD 1500 (79).

What about determining the time of night? According to Kasia Szpakowska of the University of Swansea, "Many of the royal Books of the Afterlife describe the journey of the pharaoh with the sun-god through the twelve divisions of the night" (301). Presumably, these divisions of the night were marked through the use of the water clock. The water clock came into use in Egypt Well before 1000 BC and was employed later by the Greeks and Romans (23). A water clock (Greek clepsydra, κλέπτειν *kleptein*, 'to steal'; ὕδωρ *hydor*, 'water') used a measured, regulated flow of liquid and was calibrated with a sundial. Pliny noted that the water clock replaced the sundial as the official Roman timepiece in about 150 BC. At about this same time, the Romans specified both day and night as being twelve hours long (79).

Water clocks allowed the estimation of minutes and seconds. As I noted earlier, around 250 BC Archimedes had designed gear mechanisms for water clocks to generate mechanical motions, providing the general basis for all gear design (79). "Until the perfection of the pendulum clock around 1700, the water clock was the most accurate of the world's timepieces." (ibid., pp. 58-59)

The water clock allowed the estimation of twelve segments of equal length during the night, and both the water clock and the sundial were used to estimate twelve segments of equal length during the day (23). Still, these hours varied seasonally in length between day and night and across latitudes. Of course, water clocks and sundials were not practical for use at sea.

Greeks in the 3rd century BC and, later, the Romans may have used hourglasses, but documentation is unreliable. Luitprand, a monk at the cathedral in Chartres, France, may have built an hourglass in the 8th century AD (*How Products Are Made*, Volume 5, http://www.madehow.com/Volume-5/Hourglass.html). Common use of the hourglass was recorded in Europe in the early 1300s though about 1500 (23). The timing of watches aboard European ships was supported by the hourglass during the 1500s and subsequently (186).

The idea of having hours of equal length during the day and night dates back at least to AD 127 (22, 266). Hipparchus of Niceae, working in Alexandria, proposed 24 hours for use by astronomers. Claudius Ptolemeus (about AD 90 to 168) of Alexandria divided the hour into 60 minutes, using the sexagesimal numbering system. (He also compiled a catalog of stars and constellations. These star tables provided the data used by Gregory XIII for his reform of the Julian calendar in AD 1582.) However, Ptolemeus probably based his sexagesimal division on developments by ancient Sumerians before 2000 BC, passed along by ancient Babylonians.

The sexagesimal (sixty) numbering system is used also in trigonometry and in navigation. It may have developed from an old practice, still in use, of counting in base-12. To count this way, you use the thumb to enumerate the three phalanges on each of the fingers of the same hand, *versus* the base-10 system in which you count each finger and thumb on two hands. You also display the number of iterations of twelve on the fingers and thumb of the other hand until reaching five dozens, i.e., sixty (sexagesimal).

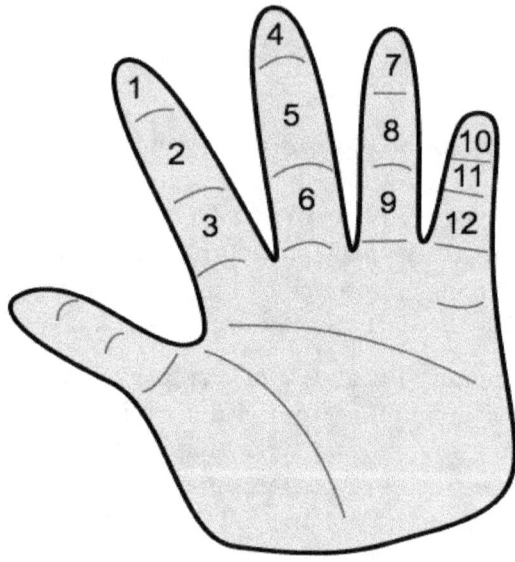

Figure 6. Base-12 counting on one hand.

Prayer times were a driving force in the development of clocks in Europe. Initially, Judaic prayers had been offered with morning and evening sacrifices (*Exodus 29:38-39, 30:6-8*) and at the third, sixth and ninth hours of the day (*Psalms 119:164*). The Christian church adopted this tradition of five prayer times. According to Harrison Cowan, in the first century AD the Romans were dividing the day into five periods, also (79). Cowan noted that Pope Sabinianus in AD 605 added two more prayer times and required church bells to be rung at these times. The bells marked the seven canonical hours. In the Middle Ages, the western (Roman) church prescribed the Liturgy of the Hours for use in monasteries, as shown in Figure 7.

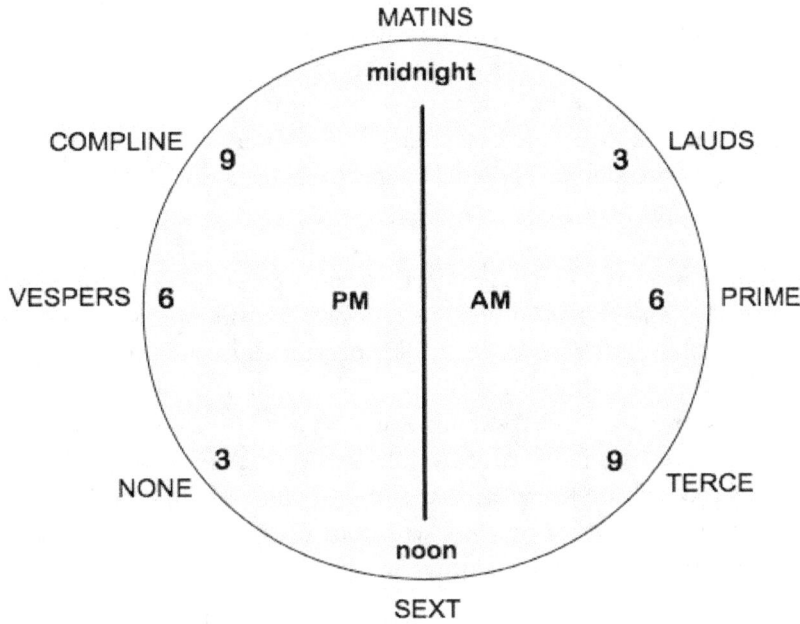

Figure 7. The Liturgy of the Hours (154).

Harrison Cowan noted a seminal change in progress concerning time measurement that occurred with the collapse of the Roman Empire and the subsequent dominance of Islam over the Persian Empire and the Mediterranean, while Europe experienced the Dark Ages.

> Manuscripts and books of the Greek philosophers, geographers, astronomers ... were carried to Bagdad and translated into Arabic... Among these was the *Almagest* of Ptolemy. Much of the learning of the ancient days was only preserved through Arabic translations... What we call Arabic numerals originated in India, and were put to use in Arabia around A.D. 800... the study of astronomy as a science was actively undertaken... Virtually unknown to European countries, greater progress was made in the sciences in remote Arabia between 800 and 900 than for a thousand years before. Jewish scholars had an important role in the universities of Bagdad, and later carried Eastern culture into Europe. (79, pp. 60-61)

Regarding developments in Islam, writer Zakaria Virk cited sources including al-Hassan and Turner and noted:

> [The] Sundial was used with a table of declination of the sun. A clepsydra was a water clock in which water flow was regulated by a siphon. A mercury clock used an astrolabe for dials, mercury flowed from compartment to compartment keeping it in perpetual motion, the astrolabe dial rotated once every 24 hours. Water clocks were built in Toledo by al- Zarqali. (314)

Also,

> Astronomical tables were used to find out eclipses, planetary motion, and trigonometric functions. Islamic Spain is best remembered for Toledan Tables produced by al-Zarqali. Christopher Columbus (1446-1506) carried a copy of astronomical tables prepared by Abraham Zacuto (1452-1515) during his voyages to the New World. Columbus fully expected to land in India, which he knew was ruled by Muslims, hence on his first voyage to America he took with him Luis de Torres, an Arabic speaking Spaniard as his interpreter. (313, 314)

As noted above, Gerbert d'Aurillac (c. AD 946 to 1003; later Pope Sylvester II, 999 to 1003), was a prolific scholar and teacher, and one who benefited from Islamic scientific advances (313). Some sources report that he invented a simple mechanism in AD 966 that rang bells at regular intervals throughout the day to call his brethren to prayer. This may have led to the practice in monasteries of the Middle Ages in which one monk would use a mechanical clock that sounded a bell. That monk then summoned the others to the prayer times observed in most monasteries. In a mechanical clock, a slowly freefalling weight would trip a lever that rang a bell. The sounding of *die Glocke* (bell, in German) gave us our first "clocks" ("clock" sounds like the German, "*Glocke*").

The number of bells at a monastery varied from four bells at dawn to one bell at noon and back to four bells at sunset (23). The intervals between bells were known as "canonical" hours. The awakening for Matins at midnight echoed the division of the long night into two sleep periods (102, 103). However, the modern Daily Office suggests prayers, for example, at the beginning of the day, mid-morning (about 0900), noon, mid-afternoon (about 1500), sunset, and at bedtime (154). Note the removal in this modern schedule of the need to arise in the middle of the night.

Even with these advances in timekeeping ideas, the ability to actually measure hours of equal length with some accuracy did not exist until the 14th century with the invention of the verge escapement (23). The "verge" was an oscillating stick (Latin, *virga*, stick) linked to a cogged wheel by two metal plates ("pallets"). Weights on either end of the stick were positioned to control the rotational speed of the stick: move the weights out and the stick oscillates more slowly; move them in and the stick oscillates more rapidly. This mechanism controlled the rate of a mechanical clock by advancing a gear train at regular intervals or "ticks." By about AD 1330, the "equal" hour had replaced the "canonical" hour (23). Pendulum driven escapements followed soon after. The clock is mentioned in the records of Richard the Second of England in 1390 (235).

The next major advancement in timekeeping was due largely to the need to measure time accurately while at sea so that longitude could be reckoned (23).

> The key to knowing how far around the world you are from home is to know, at that very moment, what time it is back home. A comparison with your local time (easily found by checking the position of the Sun) will then tell you the time difference between you and home, and thus how far round the Earth you are from home. ...
>
> The great flaw in this 'simple' theory was – how does the sailor know time back home when he is in the middle of an ocean? The obvious, and again simple answer is that he takes an accurate clock with him, which he sets to home time before leaving. All he has to do is keep it wound up and running, and he must never reset the hands throughout the voyage.
>
> This clock then provides 'home time', so if, for example, it is midday on board your ship and your 'home time' clock says that at that same moment it is midnight at home, you know immediately there is a twelve hour time-difference and you must be exactly round the other side of the world, 180 degrees of longitude from home.
>
> The principle is indeed simple, but the reality was that in the 18th century no one had ever made a clock that could suffer the great rolling and pitching of a ship and the large changes in temperature

whilst still keeping time accurately enough to be of any use. Indeed, most of the scientific community thought such a clock an impossibility. (21)

The verge and pallets were modified to become a balance wheel (Figure 8). "The vibrations of a balance [wheel] depend largely on the *mass* of the body and its *diameter*, regulated by the balance spring. The *arc* of the balance is that part of the vibration during which it is locking and unlocking the escapement..." (292).

Figure 8. Balance wheel in a cheap 1950s alarm clock, the Apollo, by Lux Mfg. Co. showing the balance spring (1) and regulator (2). (Wikipedia)

The problem here was that springs, including balance springs, unwind fast initially and slow down gradually as tension is released. Several devices were developed to deal with this inequality. One device was borrowed from a conical axle used on crossbows to manage the force applied to an arrow. The clock's mainspring was connected to a chain or piece of gut that was wound around a conical, grooved *fusee*. "Thus the fusee acts as an equalising continuous gear, compensating for the unequal torque of the mainspring throughout its run" (292). The use of the *fusee* in clocks probably originated in or before about 1430, as indicated by its presence in a table clock made during that period (128).

In the mid 1600s, Christiaan Huygens (1629 to 1695) and Robert Hooke (1635 to 1703) made great advances in the design of the balance spring. Hooke discovered the law of elasticity in about 1660. This law describes the variation of tension with extension in a spring. He applied this knowledge to the development of a balance spring for a watch. Hooke also invented a very effective anchor-shaped escapement, which was built by William Clement in about 1670.

John Harrison (1693-1776) worked on the development of the marine chronometer. One major problem that he faced was the sensitivity of the balance to variations in ambient temperature. His solution was a bimetallic sensor that adjusted the effective length of the balance spring. This scheme worked well enough to allow Harrison to meet the standards set by the 1714 Longitude Act of the British Parliament, which included a monetary Longitude Prize for a usable method for the precise determination of a ship's longitude. In 1761, Harrison won the Longitude Prize (295). His portable clock lost only five seconds on a nine-week voyage. Greater detail about this work was provided by Ruth Burkholder of the Captain Cook Society.

> John Harrison was born at Foulby in Yorkshire, the eldest son of a carpenter. He was given a watch when he was six to amuse him while in bed with smallpox. He spent hours listening to it and studying its moving parts.
>
> Harrison learned his father's trade, made extra money surveying land, read lectures on mechanics and physics, made and repaired clocks. He was a man of many skills and used these to improve on the way clocks were built. For example, he developed the "gridiron pendulum", consisting of alternating brass and steel rods assembled so that the different expansion and contraction rates cancelled each other out and there was no loss or gain of time due to temperature. Another example of his inventive genius was the "grasshopper" escapement--a control device for the step-by-step release of a clock's driving power. Being almost frictionless, it required no oiling.

In 1728 Harrison packed up full scale models of his inventions and drawings for a proposed marine clock and headed for London seeking financial assistance. He was sent to George Graham, the country's foremost horologist. He must have been impressed with Harrison for Graham personally loaned him money and told him to build a model of his marine clock.

Harrison spent seven years building No. 1--a large device weighing 72 pounds. Since pendulums had proved unreliable at sea, he contrived a system of two large brass balances connected by wires. Since the balances were opposed to each other the effect of the roll of the ship on one would be counteracted by the other. After due examination by the Board, it was deemed worthy of trial by sea. This was successful and Harrison was sent home with a subsidy from the Board of Longitude to build No. 2. This was completed while England was at war with Spain, and rather than risk having the timepiece fall into enemy hands, it was never tried at sea.

No. 3 took Harrison seventeen years to build. He must have been thinking of many improvements because he started No. 4 without submitting No. 3 for trial. His first three clocks were all of the same design—heavy box-like instruments. No. 4 was a thing of beauty--a large pocket watch, twelve centimetres in diameter, with a jewelled mechanism ...

The next step was to try it out. In November 1761, Harrison, by now sixty-eight years old, entrusted his precious watch to his son and co-worker, William. The Board of Longitude placed the clock aboard the *Deptford*, bound for Jamaica. ...

The first leg of the voyage from Spithead to Madeira generated unusual interest in the timepiece since the ship's beer had spoiled and the crew didn't want to cross the Atlantic with only water to drink. Nine days out the *Deptford's* longitude by dead-reckoning was 13 degrees 50 minutes West, but by Harrison's calculations 15 degrees 19 minutes West--a difference of 160 kilometres. Digges, sceptical but committed to testing the timepiece, kept to Harrison's course and sure enough they sighted land right where it should be. If they had altered course to go by dead-reckoning they would have missed Madeira altogether.

The landing at Jamaica was equally successful. Digges followed Harrison's calculations and arrived there three days before another ship that had left port ten days before the *Deptford*. No. 4 was taken ashore and checked against Jamaica's longitude as determined by astronomical calculations. Making a predetermined allowance for two and two-thirds seconds a day for error it was found to be five seconds slow--an error of 1.25 minutes in longitude, more than close enough for the great prize.

However the Board of Longitude decided it was just a fluke that the watch performed so well. They gave Harrison £2,500 and said that further sea trials would be necessary. So in March 1764 William set sail for the Barbados aboard the *Tartar*. Once again the timepiece performed magnificently an error of only 34.4 seconds over seven weeks or on the round trip less than a tenth of a second a day! (38)

Figure 9. John Harrison.

"Ill-used" by the Board, Harrison petitioned King George III who caused the Board to award Harrison the full prize money in 1773. "In 1776 James Cook took a copy of No. 4 with him to the Pacific. He proved it extremely useful many times over in charting and mapping the many places he visited on his second voyage" (38).

Harrison's clocks were stored away for half a century after the award of the prize. The polymath, Rupert Gould, "rediscovered the great marine timekeepers of John Harrison, corroding in store at the Royal Observatory at Greenwich. ... Although famous in his day, Harrison's life had virtually been consigned to history until Rupert Gould retold his extraordinary story in 1923" (21).

The marine clock is referred to as a chronometer, a term created in 1714 by Jeremy Thacker during the Parliament-sponsored competition. In 1766 in France, Pierre Le Roy created a revolutionary chronometer that incorporated a detent escapement, a temperature-compensated balance and an isochronous balance spring. These were the basic mechanisms needed for the marine chronometer.

Clocks could now be used at sea to measure longitude. By 1848, Greenwich time had become the standard time for England, Scotland and Wales. It was then adopted as worldwide standard time at the Prime Meridian Conference of 1884, along with the present 24 time zones of 15 degrees of longitude, each (79). Ships would anchor briefly in the River Thames at Greenwich. They used their sextant, quadrant, or similar to catch the sun at its zenith at 1200 and set the two or more ship's clocks. Then, they observed the ball at the Royal Observatory as it dropped at precisely 1300 to calibrate and synchronize their clocks with what we now call Greenwich Mean Time (GMT). Once underway, one ship's clock would be reset at noon each day when the sun reached its local zenith. This local time was then compared to GMT, still shown on the other clock. Each hour's difference from home represents 15 degrees of longitude from home. At the equator, each degree of longitude represents 60 nautical miles and 69 statute miles, tapering to zero miles at the poles.

Daylight savings time (DST), recommended by Benjamin Franklin for more efficient working hours, is used today for various purposes in various countries. However, its use is not universal and can be confusing. Flying to different U.S. states and countries in the USAF C-130E Hercules in the summers of 1967-1971, we never knew whether our destination practiced DST until we were within radio range of the airport control tower and could ask them directly.

A.-L. Breguet (1747-1823), developed the successful self-winding *perpétuelle* watch, and the first shock-protection for balance pivots. In 1810 he conceived the world's first wristwatch. In 1815, he developed an improved marine chronometer. This invention was adapted by the French Navy, and Breguet was appointed "Horloger de la Marine Royal" (watchmaker to the Royal Navy).

While the balance wheel and spring dealt well with ship's motion, Harrison's temperature sensor was only marginally acceptable. This problem was solved in about 1890 with a nickel-steel alloy, Elinvar, named for its minimal variation in elasticity at normal temperatures. The inventor was Charles Edouard Guillaume, who won the 1920 Nobel Prize for physics for this and other inventions.

I have perhaps provided too much information here about the development of timekeeping. However, it was a process that fascinated me, and this development was a necessity if seafarers were to know when the watch section was to be changed. Of course, just an hourglass and a bell might still serve that purpose! Now, on to the history of the development of maritime watchstanding.

Watchstanding at Sea

The Phoenician city-states in the eastern Mediterranean developed a maritime trading culture across the Mediterranean during the approximate period 1550 BC to 300 BC. The Phoenicians appear to have been Canaanites, and not Babylonians. Thus, their knowledge of astronomy may not have been nearly as sophisticated as that of the Babylonians. The designs of their larger vessels, biremes, did not appear to lend themselves well to the practice of carrying more than one watch section. In the 1889 *History of Phoenicia*, Canon George Rawlinson of the University of Oxford asserted the following:

> The navigation of the Phoenicians, in early times, was no doubt cautious and timid. So far from venturing out of sight of land, they usually hugged the coast, ready at any moment, if the sea or

sky threatened, to change their course and steer directly for the shore. On a shelving coast they were not at all afraid to run their ships aground, since, like the Greek vessels, they could be easily pulled up out of reach of the waves, and again pulled down and launched, when the storm was over and the sea calm once more. At first they sailed, we may be sure, only in the daytime casting anchor at nightfall, or else dragging their ships up upon the beach, and so awaiting the dawn. But after a time they grew more bold. The sea became familiar to them, the positions of coasts and islands relatively one to another better known, the character of the seasons, the signs of unsettled or settled weather, the conduct to pursue in an emergency, better apprehended. They soon began to shape the course of their vessels from headland to headland, instead of always creeping along the shore, and it was not perhaps very long before they would venture out of sight of land, if their knowledge of the weather satisfied them that the wind might be trusted to continue steady, and if they were well assured of the direction of the land that they wished to make. They took courage, moreover, to sail in the night, no less than in the daytime, when the weather was clear, guiding themselves by the stars, and particularly by the Polar star, which they discovered to be the star most nearly marking the true north. (263)

Thus, there seems to be no link from the Phoenician maritime tradition to the establishment of present maritime watchstanding plans. However, watchstanding at sea probably pre-dated the portable clock of the 1700s AD by at least a millennium.

Some historians have assumed that, until the Portuguese discovery voyages of the late 1400s, Europeans focused on navigating the Mediterranean and exploring and exploiting the coast of Africa (23). When staying close to the coast, watch sections would have been unnecessary: the crew could beach the vessel, anchor or tie up at night, as the Phoenicians may have done. If true, then full-on watchstanding would not have begun as a regular practice until the late 1400s. However, recent research disproves the assumption.

Dan Davis, a former U.S. Navy navigator turned academic, presented extensive data showing that, in addition to short journeys between nearby ports during which land was kept in sight, many documented sailing routes took ships "far out of sight of land and on passages of many days' duration" (Davis, 2009; Dan Davis, personal communication, December 2013). Wrecks explored by Bob Ballard, with whom Davis collaborates (and with whom I attended the University of California, Santa Barbara in the early 1960s; thank you Bob for putting me in contact with Dan) have helped to document these routes. Homer's *Odyssey*, contains many accounts of "open-sea sailing and sailing by the stars on extended passages" (Davis, ibid.). Unfortunately, the Greek and Roman sources tell us little about the watchstanding practices that must have occurred during those voyages. However, Davis located several relevant comments (89) and summarized to me them thusly:

> Homer refers to a system of three night watches (*Iliad* 10.251, *Odyssey* 12.312, 14.483), but says nothing of a regular 24-hour watch rotation.
>
> During the Classical Greek and Roman era (5th century BC to 4th century AD), it seems a system of four night watches was in place (Agatharchides, *De Mari Erythraeo* 106a-b; Heliodorus, *Aethiopia* 5.17.5).
>
> Roman ships making the open-sea passage to India kept several pilots (*kybernetai*) aboard, presumably in order to relieve each other at the helm on a regular watch rotation (Philostratus, *Vita Apollonius* 3.35). Carthaginian ships embarked two *kybernetai*, presumably for the same reason (Aelian, *Varia Historica* 9.40). (Dan Davis, personal communication, December 2013)

Davis also reported that:

> On top of this rather limited data, there was a common seafaring *topos*, or literary theme, of the diligent pilot standing watch all night with his eyes on the stars. It starts with Homer's *Odyssey*, Book 5, but is found repeated in Greek and Latin literature over a thousand years later. It starts with (1) a sitting helmsman fighting to maintain a vigil watch; (2) his exceptional knowledge of a short list of seafaring stars and constellations (never more than the Pleiades, Arcturus/Bootes, Orion and the Great Bear); (3) an exceptional ability to predict storms; and (4) an exceptional ability to spot land from extraordinary distances. Notable examples include Tiphys in Apollonius Rhodius' *Argonautica*, Aeneas's pilot Palinurus in Vergil's *Aeneid*, and Bato the Carthaginian pilot of Silius Italicus' *Punica*. (Dan Davis, personal communication,

December 2013)

Jumping forward a millennium or so, Harrison Cowan noted, "The seven canonical hours were the only form of 'time' observed on the ships of Columbus [late 1490s]. To keep them in order, a half-hour sandglass was turned as it emptied" (79). Someone had to remain awake at night to turn the sandglass.

As reported by Richard Phillips of the University of Tasmania, "At the time of Cook's voyage to Australia [1768-1771], regulations for watchkeeping were brief, article XXVII of the Kings Regulations and Admiralty Instructions stating that 'No person in or belonging to the fleet shall sleep upon his watch, or negligently perform the duty imposed on him, or forsake his station on pain of death ...'' (Parkin, 1997)" (253). Death is a remarkable penalty to pay for falling asleep on watch.

Watchstanding from the late 1400s through the mid-1800s may have been conducted with two watch sections: port and starboard. John Rousmaniere is a very experienced sailor and an authority on boating safety. He examined the question of early watchstanding practices (270). His source was the multi-volume reference, *Medicine and the Navy, 1200-1900*, by John J. Keevil, Jack Leonard Sagar Coulter, and Christopher Lloyd, published around 1960. Rousmaniere wrote:

> In the early 19th century, the Royal Navy suffered from exhausted crews as well as poor discipline, unwashed sailors, filthy ships, rampant disease, and high mortality (even in peacetime). At mid-century, the navy addressed these problems in a number of reforms, among them a radical change in the system of assigning and standing watches. Until then, crews were divided into two watches. The starboard watch was on deck for four hours before it was replaced by the port watch, which took the next four hours, and then back to the starboard watch, and so on. The captain, the cook, and a few others who had full-time responsibilities were exempted, but otherwise the crew spent 12 hours sailing [per day]. The navy replaced this traditional "watch-on-watch" system with another one that (reportedly) was first tried by Captain James Cook during his voyages in the Pacific [the three voyages began in 1768, 1772, and 1776, respectively]. Here, the crew was divided not into halves but into thirds. Instead of two watches there were three, and instead of 12 hours of sailing there were eight. For

every hour Jack Tar spent on watch, he had two hours off, which allowed more time for sleep, relaxation, meals, maintenance of personal gear, and cleaning the ship. Rested sailors, it turned out, were cleaner, healthier sailors. Shipboard mortality rates subsequently fell dramatically. (270)

To sum up, the available historical records indicate that, at sea, the night was divided into three watches or four watches, depending upon cultural practices. Though the water clock was available after about 1000 BC to show the procession of the night hours, it probably was impractical for use at sea. Similarly, the mechanical clock was impractical for use at sea due to temperature fluctuations and ship motion (295). Thus, until the availability of the portable clock after 1761, shipboard timekeeping was accomplished by reference to the Zodiac, when visible, and the hourglass. According to The British Horological Institute, citing input from horologist Anthony Gray, a bell was used as early as the 15th century AD to sound the time on board a ship. There is more information about ship's bells in my next chapter.

Because of the relatively small sizes of vessels, it is likely that there were only two watch sections to stand the watches from ancient times until about 1800 give or take 35 years. Even today, many submarines are crewed by only two watch sections (247, 249).

Chapter 3. "Classic" Watch Plans

Watch Plan Structure

Before I delve into the natures of various watch plans, you need to understand some basic terminology and structural characteristics of watch plans. I have addressed the structure of shiftwork systems and plans in detail in *Fundamentals of Shiftwork Scheduling* (196). In the next few paragraphs I have adapted the information from that book to focus on watchstanding at sea. I have defined here the watch system, the watch plan, and the watch schedule. It is the watch plan that is the focus of the research discussed in Section II of this book.

Watchstanding System. The information contained in a shiftwork system specification provides a generalizable approach to understanding the fundamental nature of a specific shiftwork schedule and its underlying shiftwork plan (160, 161). In the formula, below, a work day (W; or a "watch" day in the context of this ebook) is a day on which a watch starts. A free day (F) is a day on which no watch starts. The shift system sets the relative numbers of work and free periods, excluding holidays. A system is expressed as the ratio:

$$XnW:YnF$$

> where X and Y are integers (the numbers 1, 2, 3, etc.), that describe the general form of the shift system, and n is a multiplier (also an integer).

Using this approach, Knauth and his colleagues showed the logical, mathematical usefulness of the average 42-hour work week in four-section shift work plans. They also suggested two useful measures for determining the acceptability of a shift system and for comparing systems: the average number of hours worked per day (or work load) and the number of days free per year. (I stress the word average because the actual hours worked per day or week will be above and below the average.)

However, with the exception of my Close-6 plan (Plan 16, below), a watch section has no days off when a vessel is underway. Thus, the system within which nearly all underway watchstanding plans exist is a 1W:0F system. That is, a section has a watch period that starts on every calendar day while underway, and there are no days off. For a two-section system, each section stands watch an average of or an actual twelve hours per day. For a three-section system, each section stands watch an average of or an actual eight hours per day. The number of free days per year for a seafarer depends upon the amount of time that the vessel is not underway. However, those "free" days (not standing watches) may be taken up with in-port watches, training, education, administration, maintenance, and other chores.

Watchstanding Plan. The plan, or rota, determines the sequence of work (W) and free (F) periods within a watch system. The underway watchstanding plan for a two-section or three-section watch is W W W W W.... until the ship anchors or docks. There are no F days at sea except in my Close-6 plan, described later in this book (Plan 16).

The usual shiftwork notation includes 'D' (day), 'S' (swing, afternoon, evening) and 'N' (night, mid-shift) for the W periods, and 'O' (off) for the F periods. Maritime watchstanding plans use a number of nomenclatures. For example, the nomenclature for the classic two-section dogged watch, starting from midnight, is Middle watch, Morning watch, Forenoon watch, Afternoon watch, First dog watch, Last dog watch, and First watch. Nominally, in three-section plans, each watch section stands watch an average of or an actual eight hours per day and 56 hours per week, while in two-section plans each watch section stands watch an average of or an actual twelve hours per day and 84 hours per week. It is the watch plan that has been the focus of research concerning maritime watchstanding, just as the shift plan has been the focus of research about shiftwork.

Watchstanding Schedule. A watchstanding schedule builds upon the watchstanding plan, which, in turn, is one sequence within a watchstanding system. A number of interacting components make up a schedule. For watchstanding, the components are organized into the following groups:

People

Number of sections, the numbers being two, three, four, etc.

Time

<u>Watch type</u>, i.e., fixed *vs.* rotating (including dogged), slow *vs.* fast rotation.

<u>Watch length</u>, generally two-, four-, six-, eight-, and twelve-hour watches, i.e., factors of 24 hours.

<u>Watch overlap</u>, the amount of time that a watchstander must remain after his or her nominal watch end time for hand-off to the next section, and the amount of time before nominal watch start time that is required for the same purpose.

Other

<u>Watch change times</u>, especially when to begin the morning watch to allow as many watchstanders as possible to sleep well at night.

Meal times.

Work Compression. In the zero-sum nature of 24/7 operations, work compression is used to allow the expansion of continuous time off. In a zero-sum system, any gain within the system must be offset by an equal loss within the system. In watchstanding, work compression means that once you start a watch, you stay on watch for more hours. In return, you have more continuous hours off within that day. For example, in a two-section plan you might wish to stand a twelve-hour watch instead of two six-hour watches so that you may then have a continuous twelve-hour period without a watch. The trade-off is that you may become so fatigued during the twelve-hour watch that your work becomes ineffective. In maritime watchstanding, work compression has been practiced by extending the watch from the classic four-hour length to six and even eight hours. In turn, this can allow the extension of continuous periods of time off between watches from the classic eight hours to twelve and even sixteen hours.

The Classic Plans

Quite a few watch plans are in use today to support the 24/7 operation of a vessel underway. Several may be labeled as "classic" plans, at least in the views of English-speaking nations. <u>Two</u>-section, fixed-watch plans have been used extensively, with each section standing twelve hours of watch per day. Often, the two watch sections are labeled "port" and "starboard." As I noted above, watchstanding from the late 1400s through the mid-1800s may have been conducted with two watch sections: port and starboard. Some military submarines use this 12-and-12 hour plan today (249).

In recent times, when the U.S. Coast Guard sailed with only two watchstanders qualified for one position, they sometimes used a 6-and-6 hour plan (213). The start times of two-section operations are often keyed to midnight, but other start times may be used at watchstanders' or commanding officers' discretion. Also, a pair of alternating watchstanders may vary their watch lengths anywhere between two or three hours and twelve hours depending upon the fatigue and wellness of one or the other of the pair, the type of watch duty, and the pace of operations.

In the classic, <u>three</u>-section, fixed maritime watchstanding plan of the second half of the 1800s and most of the 1900s, the watchstander would stand two four-hour watches per day with an eight-hour break between watches. A large portion of a daytime eight-hour break would be used to conduct ship's work, especially maintenance.

In a classic three-section, 4-and-8 watch plans, each section stands eight hours of watch per day. The sections may be labeled "Foremast," "Mainmast," and "Mizzen," or "Red," "White," and "Blue," etc. For three-section operations there seems to be confusion over which plan is the "classic" plan: the <u>fixed</u>-watch, 4-and-8 hour plan (Plan 1, below), or the rotating <u>dogged</u> watch plan based upon this 4-and-8 plan (dogged watches are discussed below). Both seem to have been in use for more than 200 years. The fixed plan uses a four-hour watch length with watch start times beginning at midnight. Each section stands watch for eight hours per day, at the same times each day. I call this the "classic maritime" watch plan. In the brief chapter by Peter Colquhoun and Simon Folkard in *Hours of Work*, this is their "stabilized" plan (63, Figure 20.1).

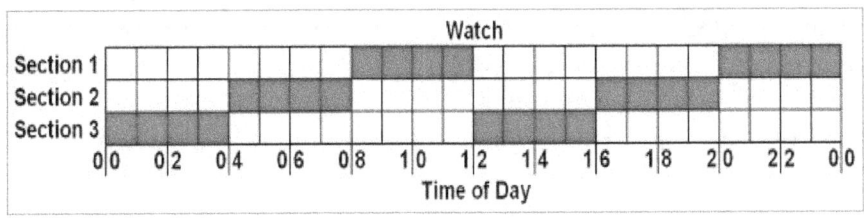

Plan 1. Three-section, fixed, 4-and-8 "classic maritime" watch plan.

Plan 1 is the first of 35 watchstanding plans described in this book. The watch periods are in red in this and my subsequent depictions of watch plans; or, if you are viewing this figure in black and white, the watch periods are the dark periods. Sleep periods are specified in some plans, and these are colored light blue in my figures.

Note that in many watch naming systems the 2000 to midnight watch is called the "first" watch (reasons discussed below). However, for clarity with respect to the day-night cycle, all of my illustrations of watch periods run from midnight on the left, through noon in the middle, to midnight on the right. Thus, a "day" in my illustrations is one calendar day.

The French Navy has apparently used a counter-clockwise rotating, three-section, 4-and-8 plan since the 17th century (304). An example is shown, below, in my section on submarine research studies reported in 2010. It is a variant of Plan 1, above.

The main problems with any 4-and-8 plan are reduced total sleep time and fragmented sleep: with only eight hours off between watches, the watchstander is never able to get the eight hours of continuous, uninterrupted sleep found by clinicians and researchers to be the average sleep need. Thus, many maritime watchstanders operate at a continual, unacceptably-low level of cognitive performance effectiveness while the vessel is underway. The French Navy variant of Plan 1 provides somewhat better sleep opportunities.

The Appendix to this book describes the methods I used to analyze quantitatively the nominal watch plans shown in this book and inlculdes tables of quantitative results. I used the Sleep, Activity, Fatigue, and Task Effectiveness (SAFTE) model and its software implementation, The Fatigue Avoidance Scheduling Tool (FAST) to model the effects of each watchstanding plan shown in this book quantitatively (48, 136–141, 198). The models covered 14 to 16 days of watchstanding and three days of recovery.

The results of my modeling are presented solely for purposes of comparison. Generally, real-world data indicate that crews sleep less than I have specified in my modeling and, thus, are even more fatigued than I have predicted.

For the classic, fixed plan (Plan 1), my modeling predicted that almost all of the time that section 1 spends in the 12 to 4 watches—<u>both</u> 12 to 4 watches--will be spent below the 90% criterion line (BCL; 90% is our normal evening sleepiness level, just before bedtime), with an average cognitive performance effectiveness of about 83% across all 12 to 4 watch periods for 15 days. By day 15 a six-hour phase delay has occurred for the body clock, as if one were traveling westbound across six time zones, with no appearance of the circadian rhythm beginning to stabilize. Post-watchstanding recovery of cognitive performance will not be complete after three ten-hour nights of sleep. This watch is known anecdotally to be fatiguing, and modeling supports the anecdotal evidence. While we expect cognitive performance to be poor during the midnight to 0400 watch, especially toward 0400, we might not expect section 1's cognitive performance to be poor, also, during the noon to 1600 watch! However, that is the prediction.

For the 4 to 8 watches, almost all watch time was predicted to be spent BCL (I used a BCL of 90% throughout my modeling efforts), with an average effectiveness of about 87% across watch periods. By day seven a two-hour phase advance of the body clock will have been completed and the circadian rhythm will have stabilized with respect to the day-night cycle. Post-watchstanding recovery will be complete after two ten-hour nights of sleep. This watch is known anecdotally to be less fatiguing than the 12 to 4 watches, and modeling supports the anecdotal evidence in terms of average cognitive effectiveness.

For the 8 to 12 watches, only about 5% of watch time will be spent BCL (below 90%), with an average effectiveness of about 97% across all watch periods. As with the 12 to 4 watches, by day 15 there will have been a five-hour phase delay of the body clock with no appearance of stabilization of the circadian rhythm, and post-watchstanding recovery will not quite be complete after three nights of ten hours of sleep per night. Thus, underway cognitive performance effectiveness will be better during this watch than during the other two watches, but fatigue will set in continually. My modeling results agree with both anecdotal and experimental descriptions of the differing effects of standing these three watch sections. Now, what happens when we introduce watch rotation by dogging the noon to 1600 watch?

Dogging the Watch

Dogging (splitting) a watch is a very old procedure in maritime operations. Classically, the 1600 to 2000 watch has been split, and this watch is called the "dog watch." There are several popular versions of the origin of the label, dog watch. However, in the light of the historical material that I developed in the preceding chapters about astronomy and the night watch, it seems most likely that the "Dog Star" (*Sirius*) version is correct.

Known by many names in many cultures, *Sirius* is the brightest star in the night sky at North American and European latitudes. During the period of the Middle Kingdom (about 2050 to 1650 BC), the Egyptian *Sothic* (their name for *Sirius*) calendar began when this star became visible just before sunrise after moving far enough away from the glare of the Sun.

The Latin name *Sirius* apparently came from the ancient Greek *Seirios*, "glowing" or "scorcher." In turn, the Greek word may have been imported from a link to the Egyptian god Osiris. The earliest recorded use of the name *Sirius* was in 7th century BC Greek poetry.

As the brightest star of *Canis Major*, the "Great Dog" constellation, Sirius is often called the "Dog Star." The ancient Greeks thought that *Sirius's* emissions could affect dogs adversely during the "dog days," the hottest days of the summer (about 3 July to 11 August), or for the Romans, the Latin, *dies caniculares*. Ancient thinking was that the excessive panting of dogs in hot weather placed them at risk of dehydration and disease. Homer, in the Iliad (just before 700 BC), describes the approach of Achilles toward Troy in these words:

> *Sirius* rises late in the dark, liquid sky
>
> On summer nights, star of stars,
>
> *Orion's Dog* they call it, brightest
>
> Of all, but an evil portent, bringing heat
>
> And fevers to suffering humanity.

When Ptolemy of Alexandria mapped the stars, he used *Sirius* as the location for the world's central meridian. Today, *Sirius* remains an important reference in celestial navigation. However, *Sirius* is not associated with the Zodiac. Thus, I cannot associate *Sirius* with my hypothesis linking the two-hour constellation appearance interval with the four-hour watch length. Instead, I suspect that the dogwatch was named because it was the evening watch period (1600 to 2000 hours) during which *Sirius* became visible in the sky. An alternative idea is an association of this watch period with a hot ambient temperature, but midday heat should be associated more with the noon to 1600 hours watch.

Here's another idea:

> Dog Watch is the name given to the 1600-1800 and the 1800-2000 watches aboard a ship. The 1600-2000 four-hour watch was originally split even to prevent men from always having to stand the same watches daily. As a result, Sailors dodge the same daily routine, hence they are dodging the watch or standing the dodge watch. In its corrupted form, dodge became dog and the procedure is referred as "dogging the watch" or standing the "dog watch." (309)

Classic Dogged Watch Plans

According to one Web source, the U.K. Royal Navy used a two-section, rotating, dogged watch plan traditionally (222). This plan uses four-hour watch lengths beginning at midnight (Plan 2, below). The 1600 to 2000 watch is dogged (split), and the plan has a two-day cycle length. A section stands watch for a total of 14 hours on one day, and ten hours on the next day, on the inverse hours from the previous day. The section oscillates between these two one-day plans. Work compression on the 14-hour day (Day 1) allows less watch time and more time off during the ten-hour day (Day 2).

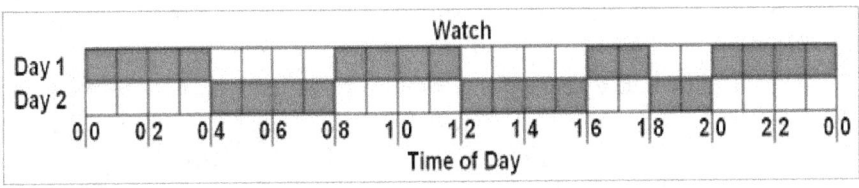

> **Plan 2.** Two-section, dogged, classic U.K. Royal Navy watch plan. Read each line left to right, then drop down a line for the next day.

The main problem with any two-section plan is the lack of uninterrupted nocturnal sleep for the section that works in the midnight to dawn period. About half of the maritime watchstanders in a two-section plan may operate at a continual, unacceptably-low level of cognitive performance effectiveness while the vessel is underway.

In Plan 2, my modeling indicated that average cognitive performance effectiveness across all (day and night) watch periods would be about 85%, with about 70% of all watch time spent BCL. The body clock would phase delay slowly about 0.5 hour across 14 days. Two recovery nights would be needed.

Apparently, the Swedish Navy has used a different approach to the dogged two-section watch plan (54). In this two-day cycle, watch lengths of 4-4-5-6-5 are used (Plan 3). The odd number of five watches per 24 hours causes watch rotation.

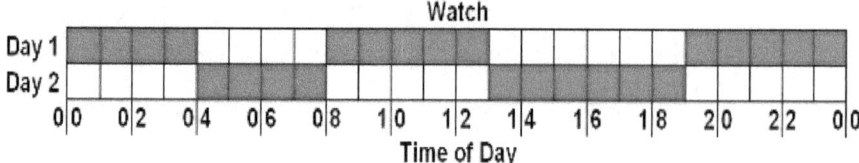

Plan 3. Two-section, rotating, Swedish Navy watch plan. Read each line left to right, then drop down a line for the next day.

My modeling suggested that the fatigue effects of Plan 3 were worse than those of Plan 2. Average cognitive performance effectiveness across all watch periods was predicted to be about 77%, with about 91% of all watch time spent BCL. The body clock would phase delay about three hours across 14 days with no circadian stabilization in sight. More than three recovery nights of ten hours of sleep would be needed.

In a three-section watch plan, dogging a watch causes watch rotation, also. With three watch sections, the cycle length of the plan is, necessarily, three days. Here is an example of dogging the classic maritime plan (Plan 1) during the 1600 to 2000 hours watch (Plan 4, below). I call this the "classic dogged maritime watch plan." It is also the "Rotating I" plan used by Colquhoun and Folkard (63, Figure 20.1).

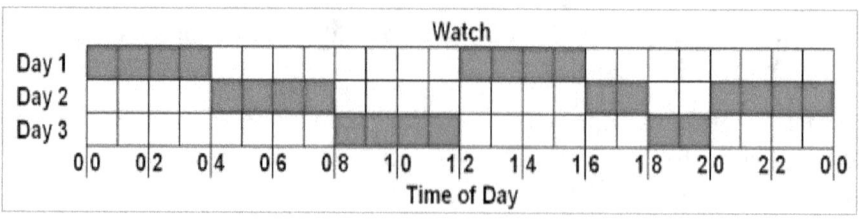

Plan 4. Three-section, 4-and-8, classic dogged maritime watch plan, dogged in the 1600 to 2000 watch. Read each line left to right, then drop down a line for the next day.

In Plan 4, a given watch section stands watch for eight hours, ten hours, and six hours on three successive days, respectively. Except at the juncture between the end of the cycle and its beginning, watch rotation is forward on the clock, or clockwise: each daily pair of four-hour watches starts later on the clock than the preceding daily pair of watches. Work compression on day 2 allows fewer hours of watch on day 3. However, there is compression of time off at the end of day 3. Some names for the watches in the classic dogged plan are (18,84):

- 0000 to 0400: Midwatch
- 0400 to 0800: Morning watch
- 0800 to 1200: Forenoon watch
- 1200 to 1600: Afternoon watch
- 1600 to 1800: First dog watch
- 1800 to 2000: Last (or second) dog watch
- 2000 to 2400: Evening (or first) watch

In some non-classic cases, the 1200 to 1600 watch may be dogged instead of the 1600 to 2000 watch.

In Plan 4, my modeling suggested that <u>all three</u> sections will be fatigued. The average effectiveness across all watches will be about 84% with about 80% of all watch time spent BCL, much like the 12 to 4 watches in classic Plan 1. There will be a stabilized one-hour circadian phase delay of the body clock by day 4, and the body clock will oscillate very slightly each three days after that. Thus, while dogging may enhance circadian stability compared to the fixed plan, cognitive performance effectiveness will be generally poorer with this dogging approach, being reduced across all three sections such that the average predicted cognitive performance will be at about the level of the 12 to 4 watch section in the classic fixed Plan 1. This similarity may help explain why there are mixed anecdotal opinions about dogging the classic maritime watch plan.

The modern French Navy has used a different dogging method to induce watch rotation. One four-hour watch is extended to five hours and another is shortened to three hours (see Plan 24, below).

Ship's Bells

According to the U.S. Navy Library,

> One of the earliest recorded mentions of the shipboard bell was on the British ship *Grace Dieu* about 1485. Some ten years later an inventory of the English ship *Regent* reveals that this ship carried two "wache bells."

> Before the advent of the chronometer time at sea was measured by the trickle of sand through a half - hour glass. One of the ship's boys had the duty of watching the glass and turning it when the sand had run out. When he turned the glass, he struck the bell as a signal that he had performed this vital function. From this ringing of the bell as the glass was turned evolved the tradition of striking the bell once at the end of the first half hour of a four hour watch, twice after the first hour, etc., until eight bells marked the end of the four hour watch. The process was repeated for the succeeding watches. (310)

I obtained the following information from an archived Web page of The British Horological Institute. In turn, the Institute cited input from horologist Anthony Gray. I adapted my table of a ship's bell schedule from this web page (Figure 10).

> As early as the 15th Century a bell was used to sound the time on board a ship. The bell was rung every half hour of the four-hour watch. A 24-hour day was divided into six four-hour watches, except the dog watch (1600 to 2000 hours) which could be divided into two, two-hour watches to allow for the taking of the evening meal.

> After a mutiny on the *Nore* in 1797, British ships modified the bell system on the dog watch (1600 - 1800) so that the mutiny signal of five bells was never again struck.

The 1797 date cited here conflicts somewhat with the statements by Rousmaniere, cited above, concerning the use of only two sections as late as the early 1800s. This bell system implies the use of three sections as early as about 1770. There may have been a transition period of 30 or even 60 years to move from two- to three-section operations, driven by limits in onboard accommodations and/or by the unwillingness of some captains to try the "new" system. Of course, two-section crews continue to operate in submarines today, as mentioned above.

Watch	Standard	British Royal & Merchant Navy Clocks after 1797
Middle (Graveyard) Watch	00:30 1 bell 01:00 2 bells 01:30 2 bells, pause, 1 bell 02:00 2 bells, pause, 2 bells 02:30 2 bells, pause, 2 bells, pause, 1 bell 03:00 2 bells, pause, 2 bells, pause, 2 bells 03:30 2 bells, pause, 2 bells, pause, 2 bells, pause, 1 bell 04:00 2 bells, pause, 2 bells, pause, 2 bells, pause, 2 bells	Same as Standard
Morning Watch	04:30 - 08:00 Same as previous watch	Same as Standard
Forenoon Watch	08:30 - 12:00 Same as previous watch	Same as Standard
Afternoon Watch	12:30 - 16:00 Same as previous watch	Same as Standard
Dog Watch	16:30 - 20:00 Same as previous watch	16:30 1 bell 17:00 2 bells 17:30 2 bells, pause, 1 bell 18:00 2 bells, pause, 2 bells 18:30 1 bell 19:00 2 bells 19:30 2 bells, pause, 1 bell 20:00 2 bells, pause, 2 bells, pause, 2 bells, pause, 2 bells
First Watch	20:30 - 00:00 Same as previous watch	Same as Standard

Figure 10. Ship's bells, per The British Horological Institute. Note that the first bell occurs thirty minutes into a watch, and that the end of one watch and beginning of the next watch is marked by eight bells.

According to the Centennial edition of the U.S. Navy's *The Bluejacket's Manual* (84),

A system that used a half-hour sand-glass and the ship's bell was created and used for hundreds of years [at least since the early 1500s; see below]. At the beginning of a watch, the sand-glass was turned over to start it running. As soon as it ran out, the watchstanders knew the first half-hour had passed, so they rang the ship's bell once and immediately turned the sand-glass over to start the second half-hour. ... When the sand ran out the second time, the watchstanders rang the ship's bell twice. They continued this until eight bells had been rung (representing the passage of four hours or one complete watch.) ...

Today, because bells are rung more out of tradition than for real function, they are not normally rung between taps and reveille (normal sleeping hours for Sailors not on watch) ... (pp. 334-335)

The bell schedule cited in *The Blujackets' Manual* was the standard schedule shown by The British Horological Institute, above.

Naming of the "First" Watch

Note that the "first" watch, above, began at 2000. The tendency to call the 2000 to midnight watch the first watch depended most likely upon Judeo-Christian adherence to the account of Creation in Scripture. Citing Moses again in *Genesis*, "And there was evening and there was morning, the first day" (*ESV*), we see that evening preceded morning in each day of the Creation. Thus, the tradition developed that the day starts at sundown, especially for Jewish religious holidays.

K.M. Mathew noted:

> During the voyage of Castro to India on board 'nau Grifo' in 1538, the crew kept watch on board. In order to felicitate the division of the day into quarters of four hours each, Castro gave different names to each quarter of four hours: *prima* meant from 8 P.M. to mid-night, *madorra* meant from mid-night to 4 A.M., and *alva* meant from 4 A.M. to 8 A.M. Castro often referred to this *alva* in his first *roteiro* [navigational route description] of 1538. Each quarter was determined by sand clock of half an hour so that one half hour of a quarter was called one clock and so each quarter consisted of eight clocks. (186, p. 27)

My friend in Portugal, André Pimenta, informed me of the meanings of the ancient words used above.

Madorra means state of sleep, sleepiness caused in some cases by illness or diseases.

Alva means first light of day that produces the horizon.

Note (1) the assignment of "first" (*prima*) to the 2000 to midnight watch, (2) the use of the sand clock, and (3) the agreement of the naming of the half-hours with the description of the ship's bells from the 1700s, above.

Alternatively, references quoted by 7th day scholar Frederick Steed (1922-2007) in an undated essay, "When Does the Day Begin?" (http://www.HouseofSteed.com), suggested that ancient diurnal activities such as Temple worship and farming may have been reckoned on a morning-to-morning basis. Similarly and much later, the first prayer time, "Prime," in the Middle Ages was at 0600, as I noted above.

However, throughout the millennia many religious holidays were governed by a lunar calendar. Since the period of a "day" was associated with observations of the moon's cycles, a day was reckoned from evening to evening. For example, we still celebrate holiday "eves" in the modern Christian church calendar. Perhaps the general superstition and religiosity of sailors encouraged adherence to the more religious approach, causing them to start their new day in the evening.

Standing during the Watch

According to a U.S. Navy source, the use of four-hour watches was "based upon normal shipboard conditions and has been traditional for centuries" (84). This is probably true, considering the two-section, rotating, dogged watch plan used traditionally by the U.K. Royal navy (Plan 2, above). The three-section four-hour watch probably did not come into common use until the early or even mid-1800s. Also, "Four hours is the standard because that is widely accepted as the optimum time for a person to carry out the duties associated with operating a ship without suffering the dangerous effects of fatigue." (84, p. 334).

It is my understanding that there was a maritime tradition of literally "standing" a watch for four hours while a ship was underway, and that only in the later 20th century were provisions made for some watchstanders to sit. However, I have not found explicit statements supporting this view. Sleeping on watch was prohibited, of course, in the earliest U.S. Navy regulations:

> ART. 33. If any person shall sleep upon his watch, or negligently perform the duty which shall be enjoined him to do, or forsake his station, he shall suffer such punishment as a court-martial shall think proper to inflict, according to the nature of his offence. (311)

Standing was implied for the senior watch officer in the British Navy of 1794:

> The lieutenant, who commands the watch at sea ... During the night-watch, he occasionally visits the lower decks, or send thither a careful officer, to see that the proper centinels are at their duty, and that there is no disorder amongst the men; no tobacco smoaked between decks, nor any fire or candles burning there, except the lights which are in lanthorns, under the care of a proper watch, for particular purposes. He is expected to be always upon deck in his watch... (305)

And for the

> CORPORAL of a ship of war ... to walk frequently down into the lower decks in his watch, to see that there are no lights but such as are under the charge of proper centinels, which he is to see placed, &c. (Truxtum, op. cit.)

Even later, in the 1930 Articles of the U.S. Navy,

> Article 4. The punishment of death, or such other punishment as a court martial may adjudge, may be inflicted on any person in the naval service -- ... [who]
>
> [S]leeps upon his watch;
>
> Or leaves his station before being regularly relieved; ... (36)

I suspect that standing up throughout a watch was judged to be preferable to this latter punishment. Maybe this is why the tradition or practice of standing throughout the watch came into being.

Additionally, the tradition or practice of remaining standing throughout a watch may have evolved from the observation that standing mediates greater alertness than sitting, an observation confirmed eventually by measurements of the cortical arousal level of the brain and performance on an un-alerted reaction time test (44, 45).

To this day at our U.S. military academies cadets use the alerting effect of standing. The time demands of taking about six solid courses each semester, participating in intercollegiate or intramural athletics, participating in military activities, and studying to sustain a passing grade point average leaves a cadet little time for sleeping. In 2000, I supervised a project by USAF Academy (USAFA) seniors in which we surveyed cadets about their sleep. We found that 80% of the cadets surveyed acquired about 3.7 to 6.85 hours of sleep per night, and that 10% of cadets were acquiring less than about 3.7 hours per night. The average daily nap length (i.e., sleep not acquired at night) was about 1.7 hours, or about 24% of their total daily sleep.

The cadets are sleepy kids much of the time. When they perceive that are starting to doze off in class, they are <u>required</u> to move quietly to the back of the room and stand against the wall to avoid falling asleep. My observations when teaching classes at USAFA were that the practice was quite effective. Fortunately, falling asleep in class was not punishable by death. We lost enough cadets through non-fatal attrition.

The four-hour watch length limit may have evolved in part from a perceived upper limit for the length of time that a watchstander could remain standing twice a day, as opposed to an eight-hour continuous watch length. This limit may have been associated with subjective perceptions of venous pooling/venous return, foot pain and/or lower back pain.

Venous Return and Venous Pooling. After arteries transport oxygenated blood from the heart to the feet, veins serve to return less-oxygenated blood from the feet back to the heart. The femoral veins, popliteal vein and deep veins of the calf carry about 90% of that returning blood. These are the veins that rarely may suffer clotting (deep vein thrombosis) in some people due to prolonged sitting during long commercial airline flights.

Very little of arterial blood pressure is left after the blood departs the capillaries. Instead of 80 to 120 mmHg, the pressure in the veins may be only 10 mmHg. Additionally, the gravitational force on a column of water or blood creates a downward force known as hydrostatic pressure. When you stand upright, hydrostatic pressure assists downward arterial flow to the feet, but impedes upward venous return from the feet back to the right side of the heart. The low pressure in the leg vein is insufficient to get blood back up to the heart. Venous return from the legs to the right side of the heart receives two kinds of assistance. First, valves occur at regular intervals within the leg veins. These valves allow the blood to move only upwards in the deep veins.

Second, the deep veins run within the calf muscle.

> Each time the calf muscle contracts, it squeezes the deep veins, emptying these veins of blood. Because of the one-way venous valves, the blood can only go "up" or back toward the heart. Thus, the calf muscles and the veins within them form a calf muscle pump. The calf muscle pump provides upward propulsion to overcome the downward hydrostatic pressure and send venous blood back to the heart.
>
> The calf pump is idle when you stand still or sit with your legs in a dependent position. In these positions, gravity causes blood to pool in the lower legs. This pooling of blood makes your legs feel tired, heavy and achy. Prolonged standing still or sitting with legs dependant are therefore bad for your veins. Exercise and leg elevation are good for your veins because they prevent pooling of blood. (website of The Vein Center of North Texas)

Fortunately, the deep veins tend to be resistant to unwanted dilation and valvular failure. Thus, we would expect major venous occlusion problems in the legs to occur only in older mariners. In younger mariners, there might be a small, non-threatening drop in blood pressure associated with prolonged standing (163, 233).

Foot Pain. Differences in foot pain between morning and afternoon during two hours of standing are somewhat small (about 10%) and variable, but the difference is certainly present (163). I suspect that the cessation of standing, followed by an extended period of rest or of varied duties, after a half-day (i.e., four hours) of work that requires standing may prevent a 10% increase in foot pain. Across many days of watchstanding, that could add up to a very meaningful effect.

Back Pain. By measuring the electrical activity in the *gluteus medius* muscles, located one on each side within the pelvis, one may predict reasonably well who among people standing continually for two hours will or will not perceive pain in the lower back (231). This research by the Department of Kinesiology of the University of Waterloo in Ontario,

Canada was conducted with folks who were asymptomatic for lower back pain. The change in muscle activity was only about 10%, but even this small amount of change was pretty accurate in predicting the occurrence or non-occurrence of back pain.

David Antle and Julie Côté of McGill University in Montreal and the Jewish Rehabilitation Hospital in Laval, Canada, conducted an integrative examination of "the impact and interactions between vascular, muscular and balance outcomes" (5). They measured blood flows, ankle mean arterial pressure, muscle activities, balance, and leg and back discomfort in volunteers performing a repetitive box-folding task. Their results suggested that "the origin of standing-related lower limb discomfort is likely vascular in origin, whereas back discomfort is likely multifactorial, involving muscular, vascular and postural control variables."

Interestingly, when people are placed upon a sloped surface of plus or minus 16 degrees in the same kind of test, there are sharply different responses of the *gluteus medius* muscles, and "an exit survey satisfaction rating with 87.5% indicating that they would use the sloped surface if they were in an occupational setting that required prolonged standing work" (230). At sea, of course, the deck tends to slope one way or another and to oscillate. I experienced +/- 21 degrees rolling on a modern 210-ft USCG WMEC cutter off the Washington-Oregon coast during my 1990s study of crew fatigue (213).

Thus, my working hypotheses are that (1) the prevention of the 10% change in muscle activity (above) allowed by limiting watch length and splitting the eight hours of watch into two periods was a factor in the four-hour watch length limit, and that this was learned through experience; and that (2) the deck angle and ship oscillation may have been protective in terms of both venous return and back pain prevention.

Recently, Nita Lewis Shattuck, Panagiotis Matsangas, John Moore, and Laura Wegemann of the U.S. naval Postgraduate School (NPS) published a report that addresses some of the musculoskeletal issues in maritime operations (284). 767 crewmembers of a U.S. Navy aircraft carrier completed an anonymous survey that covered "demographics, exercise frequency, average sleep duration, caffeine consumption, the Epworth Sleepiness Scale (ESS), the Nordic Musculoskeletal Symptoms Survey, and the Fatigue Severity Scale (FSS)."

Two-thirds of the sample was male. The average age was 25.4 ± 5.94 years; ranging from 18 to 49. 43% of the sample was overweight and 7% obese. ... Approximately 58% of the respondents reported at least one MSK [musculoskeletal] symptom in the last 12 months, 44% reported at least one symptom in the last seven days, and 20.4% reported that MSK symptoms prevented them from carrying out their normal activities. Regarding the 12-month prevalence, the lower back (39.5%) and knees (33.6%) were the two body parts most frequently reported for MSK symptoms. Symptoms in these two body parts were also the most frequently reported as preventing participants from normal activities. Older crewmembers were more likely to report MSK symptoms, and females reported more MSK symptoms than males. The occurrence of MSK symptoms was associated with elevated fatigue levels and excessive daytime sleepiness. Compared to crewmembers without MSK symptoms, crewmembers with MSK symptoms are more likely to report elevated daytime sleepiness ($g=0.26 - 0.39$), increased levels of fatigue ($g=0.54 - 0.59$) and are more likely to use sleep-promoting medications. Crewmembers reporting MSK symptoms were more likely to consume caffeinated beverages. (Abstract)

Keeping Watch during the Watch

To keep watch, one must remain vigilant. Vigilance is defined in the human factors research community as the ability to detect the occurrence of a rare but important signal embedded in a background of numerous and similar, but unimportant events (211). Think of a security person watching a dozen video screens day after day and hour after hour, each screen showing lots of store customers (the "events"), looking for the rare shoplifter (the "signal"). The study of human vigilance performance dates back to World War II at the Applied Psychology Unit (APU) of Cambridge University, England. Sir Frederick Bartlett directed the APU from 1944-1951. He was succeeded by Norman Mackworth, who had by then defined the field of human vigilance research as it exists today (199).

The investigations initiated at the APU concerned the ability of the human operator to sustain attention to a task. This work was conducted in response to difficulties with sustaining attention reported by wartime radar operators. Mackworth's classic studies of signal detection in tasks simulating those of radar operators provided a paradigm for much subsequent vigilance research. His Clock Test was used to establish one of the fundamental findings human behavior: the vigilance decrement, in which signal detection accuracy decreases about 30% across 30 minutes on the task (179).

This classic vigilance decrement suggested a continuous decline in perceptual sensitivity. However, a different kind of performance impairment was shown by Donald Broadbent and others at the APU in studies of reaction time tasks. People performing monotonous tasks tend increasingly to produce unusually slow responses times that have been termed 'blocks' or 'gaps', that may reflect a lapse in attention or even a "microsleep." Reaction time tasks also provided the basis for studies of interactions of stressors, and evidence that, in some instances, fatigue might be countered by agents that elevated arousal, such as amphetamine (28).

Some of the premier applied work in this field was performed and reviewed by Robert R. Mackie and his colleagues (34, 176, 177, 191–193, 211). Much of that work focused on sonar operations in the U.S. Navy and was classified. Roy Davies and colleagues reviewed the numerous factors that influence vigilance decrement. These include task demand factors such as memory load and event rate, variables that influence motivation such as performance feedback, and adverse environmental conditions. Joel Warm and colleagues concluded that vigilance requires hard mental work and is stressful (317).

Contemporary studies of vigilance are most often based on measurement of 'misses' and 'false positive' errors in line with the theory of signal detection (119). However, an alternative approach must be noted here. Robert Wilkinson developed a simple cassette-tape-based, ten-minute test of "arousal" and "continuous, concentrated attention," called the unprepared simple reaction time test (USRT) (326).

Subsequently, David Dinges introduced a solid-state version of the USRT called the Psychomotor Vigilance Task (PVT), which has been used in several underway studies of watchstanding (e.g., 94, 173). However, the USRT/PVT has a signal probability of 1.0, unlike many classic vigilance tests that have signal probabilities of less than 0.05 (211). While the USRT/PVT is sensitive to fatigue (especially sleepiness), it is difficult to classify it as a "vigilance" task because it is such a short task and its signals are not embedded in a background of high-frequency, non-meaningful events. Thus, though the USRT/PVT addresses some aspects of sustained attention, and captures errors of omission in the form of lapses, it fails to address the visual search and decision-making components of operational vigilance performance.

One must also note the seminal contribution of the late John Stern of Washington University in St Louis (297). Stern's revelation of the information content present in the conformation of the endogenous eyeblink was at least as meaningful as Mackworth's characterization of the vigilance decrement and the description of the circasemidian rhythm in body temperature by Colquhoun and Blake. Stern and colleagues indicated that:

> The endogenous eyeblink is an ocular response with a characteristic rate and waveform, and is coordinated centrally with cognitive events. ... its production is coordinated with stimulus occurrence ..., with motor response production, and with other oculomotor events. Parameters of the endogeous blink are sensitive indicants of cognitive activity. The allocation of attentional resources, the subject's level of activation, and the effects of accumulated time-on-task are all evidenced by variations in endogenous blinking. (Stern et al., ibid, p. 31)

A useful measure derived from Stern's work is known as PERCLOS, referring to the percent closure of the eyelid (95, 107, 125). This measure may be obtained remotely with video devices. As far as I know, it has not been used in maritime research to date. I used a different oculgraphic device, the FIT-2000 in my study of USCG cutter crew fatigue (213, 208, 272)

SECTION II. RESEARCH

In this Section I address research concerning the issues associated with both surface and then submarine watchstanding practices. There is also a transitional chapter between the surface and submarine sections that deals with the difference between deck operations and belowdecks work, with and without exposure to sunlight, respectively. Most of the available research focuses on the three-section solution to operating a vessel 24/7 while underway. However, there are a few studies that relate to two-section solutions, also. The chapter topics are as follow:

> - An annotated bibliography of research aimed mainly at surface operations or conducted in generic laboratory simulations, in chronological order (chapters 4-7).
>
> - A consideration of studies about the effects of controlled lighting belowdecks and in submarines (chapter 8).
>
> - An annotated bibliography of research concerning submarine operations, in chronological order (chapters 9-10).

These chapters are adapted and expanded extensively from my annotated bibliography of shiftwork research (200). This present review of the relevant research literature is certainly not fully comprehensive. I have no information from modern major maritime powers such as China and Russia. Additionally, I am sure that there are many government reports from the United States and other English-speaking countries, and reports written in languages that I cannot translate that I've either missed or that are not available to the public. I would be pleased to receive electronic copies of such reports from readers and include them in a future edition of this work.

Before discussing research relevant to maritime watchstanding plans, I have provided primers on sleep physiology and circadian rhythms.

Sleep Primer

This primer is borrowed from my *Fundamentals* manual (196). Sleep is not a passive or vegetative state, as many assume. It is generated by complex activities in the brain. To emphasize this point, consider several complexities of sleep physiology.

- Three kinds of sleep are identifiable with scalp electrodes (electroencephalogram, EEG). They are slow-wave sleep (SWS; sleep stages 3 and 4), rapid-eye-movement (REM) sleep, and stage 2 sleep. Generally, about half of a good-quality, normal night of sleep is spent in stage 2, while SWS and REM each occupy somewhat less than one-quarter of the night, and the remainder of the night is made up of drowsiness (stage 1) and wakefulness (stage 0) (264, 327).

- The occurrence of SWS is associated with a release of growth hormone from the brain (319). Thus, we assume that SWS is a period during which some repair of muscle and nerve cells occurs, following their use during the preceding waking period(s).

- The occurrence of sleep is associated with memory consolidation (24, 25).

- The proportions of sleep time spent in SWS and REM sleep depend quite a bit upon the degree and nature of sleep debt.

- The sleep stages mentioned above occur in an orderly manner with a 90-min cycle (Figure 11). When you first fall asleep in the evening and sleep through the night, you pass quickly through stages 1 and 2 into much-needed SWS. Eventually, you return to stage 2 and may generate some REM sleep. This cycle takes about 90 minutes. The cycle is repeated throughout the night with relatively less SWS and relatively more REM sleep as morning approaches. Five 90-minute cycles occur across 7½ hours. These five cycles plus some falling-asleep and waking-up times lead you to spend about eight hours in bed. A rough depiction of the 90-minute sleep cycle is illustrated on the next page.

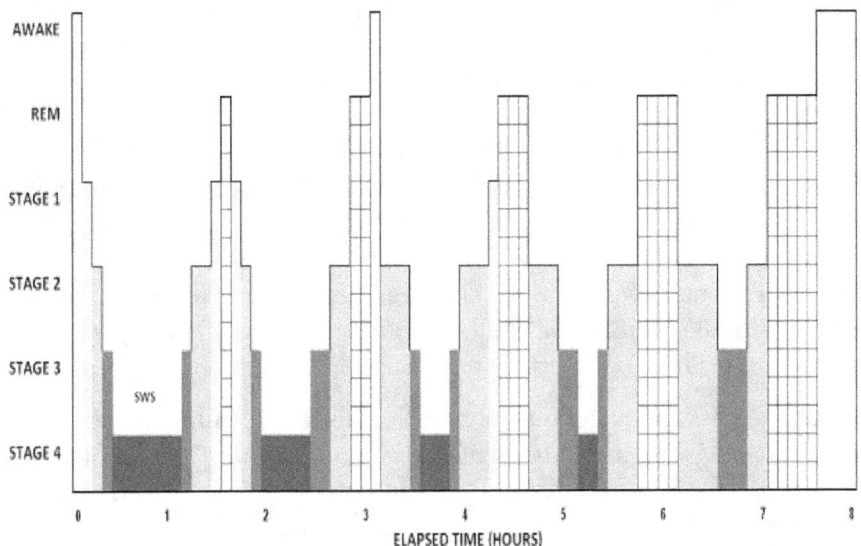

Figure 11. Rough approximation of an eight-hour sleep histogram, showing the 90-minute cycle of sleep stages that occurs throughout the night (author).

Sleep Regulation. The brain regulates the amount of sleep that we need. This regulation operates somewhat like the thermostat on a furnace or air conditioner, generating a condition known as *homeostasis*: the ability or tendency of an organism or cell to maintain internal equilibrium by adjusting its physiological processes (American Heritage Dictionary). A thermostat triggers heating or cooling when a room temperature exceeds a defined range of temperature, driving the room temperature back toward a given "set point." Similarly, in the absence of sleep pathologies, when we are too sleepy, our sleep homeostat drives us to fall asleep; and when we have recovered enough, our sleep homeostat drives us to awaken.

Sleep Drive. The need for sleep is a physiological drive, much like the drives for food and water. As with insufficient food and water intake, insufficient sleep leads to irritability and, if continued, to health problems. One may argue that sleep drive is even stronger than the drives to eat and drink. I may starve myself to death voluntarily or refuse to drink and then die of dehydration. However, continued sleep deprivation eventually causes each of us to fall asleep, initiating automatic recovery.

"[N]o one gets used to not getting enough sleep. They might be able to do it, but they never overcome the drive for sleep or the consequences that invariably follow sleep restriction." (40). Sleep debt has been compared by David Morgan to borrowing from a bank (220). People who sleep less than 8 hours per 24 hours are taking "little 'loans' from their sleep banker." Morgan cautioned that "You know that your dangerously moody sleep banker may call in the loan when you are driving at 79 miles per hour on the freeway." You should "deposit eight hours in your sleep bank every day." In fact, the U.S. National Sleep Foundation has integrated the results of decades of sleep research and observed that the normal range of sleep need for adults (about ages 25 to 70 years) is seven to nine hours per night.

Sleep Inertia. Sleep inertia occurs normally in the morning after awakening. It is a grogginess that usually lasts only about five minutes but may last up to 15 or 30 minutes in a person with a large, previous sleep debt. Though sleep inertia may not be detected as a problem after napping (Driskell & Mullen, *Human Factors* 47(2), 360-377, 2005), it is possible that if the deeper stages of sleep occur during a nap, then this same sleep inertia may occur. Thus, at least 15 to 30 minutes should be allowed after a nap to allow sleep inertia to dissipate before performing safety-sensitive jobs.

Effects of Inadequate Sleep. Humans have specific physiological and psychological requirements for getting adequate sleep. Everyone knows how it feels to get too little sleep. Sleepiness may be defined as an untimely desire to sleep and/or difficulty staying awake when wakefulness is required. Mental fatigue includes many symptoms, such as malaise, impairment of mood, memory impairment, slowed response time and impaired vigilance. Both sleepiness and mental fatigue are caused primarily by lack of sleep. When we do not get adequate sleep, we experience excessive mental fatigue during the time we are awake, also called excessive daytime sleepiness (EDS). EDS often affects our ability to perform our jobs safely. Repeated sleep loss of even one or two hours per night will eventually degrade alertness and mental performance significantly, with greater effects for greater amounts of sleep debt. The effects of one night of sleep deprivation, for example not sleeping for 40 continuous hours may still be detected in performance levels after five nights of sleeping for six hours per night (143).

We measure both mood and mental performance during laboratory and field investigations of the effects of fatigue (178, 201). Generally, fatigue-induced impairments of mood and of mental performance do not occur exactly in parallel. For example, after a night of sleep deprivation we may experience mood elevation when the sun comes up, but our mental performance may still be insidiously impaired. Conversely, we may be convinced, again insidiously, that our mental performance is adequate even though we are feeling the malaise of fatigue. This latter situation leads to single-vehicle, run-off-the-road traffic accidents in which the driver falls asleep at the wheel. This latter situation also underscores the need to create fatigue-reduction strategies when we are well-rested, not when we are already fatigued. Fatigued people tend to make "stupid" decisions.

Shift Work Sleep Disorder (SWSD). SWSD is described in Code 307.45-1 in the International Classification of Sleep Disorders (ICSD) of the American Academy of Sleep Medicine (AASM) as a circadian sleep disorder. The essential features of SWSD are "symptoms of insomnia or excessive sleepiness that occur as transient phenomena in relation to work schedules."

If we lose sleep across successive nights, the multiple deficits cause an accumulation of sleep debt and cumulative fatigue. Frequently, we tend to gain some recovery sleep over our weekends or non-work days. However, recuperation from cumulative sleep debt requires getting more sleep per night than the individual typically needs and for at least several nights (15, 143, 271).

Circadian Rhythm Primer

This primer is also borrowed from my *Fundamentals* manual (196). Circadian (sir kay' dee un) effects are normal, inherent, unavoidable, 24-hour rhythms in human cognitive and physical performance (237). The word comes from the Latin roots, *circa*, about, and *dia*, a day: the rhythm completes one full cycle in about 24 hours. Body temperature and many hormones display circadian rhythms. The rhythms are driven by a clock process located in the suprachiasmatic nucleus (SCN) of the brain's hypothalamus; the SCN is the "body clock."

Many, but not all, of these circadian rhythms oscillate between a high point late in the day to a low point in the pre-dawn hours. Both mental and physical performance follow this pattern with a trough at abut 4 a.m. local time. Mental performance usually displays a peak-to-trough amplitude of about 10% of its average value in folks with normal work and sleep schedules. Physical performance displays about a 5% peak-to-trough daily change.

Human circadian rhythms are actually slightly longer than one cycle per day, but are usually slaved, or entrained, to exactly one cycle per day by external time cues called *Zeitgebers* (tsite' gay bers, German--literally "time givers"). The main cue is the 24-hour daylight-darkness cycle. Minor cues include social activities and meals, but their effects on the body clock are so small as to be difficult to measure.

There are interactions among the body clock, the daylight-darkness cycle and the brain's ability to generate sleep (165). Daytime sleep duration is impaired during night work (159, 302). In night work, circadian effects cause, among other things, feelings of malaise and fatigue during the midnight-to-dawn period on the body clock. This is the period when sleep drive and sleepiness are highest and body temperature and alertness are lowest.

Jet Lag. Jet lag is the feeling of malaise and fatigue that accompanies a time zone change that is faster than about one time zone per day. Thus, it is hardly noticeable when traveling east or west by highway, rail or ship, but quite noticeable when traveling east or west by commercial jet. Jet lag occurs during the period of re-synchronization of the body's circadian rhythms to the day-night cycle of the new time zone. Rapid time zones changes of three hours and more are detectable in mood (318).

Shift Lag. Shift lag is the feeling of malaise and fatigue that accompanies a change from day work to night work and *vice versa*. Shift lag occurs during the period of attempted re-synchronization of circadian rhythms to unchanged external day-night cues. Compared to jet lag, the attempt to re-synchronize to a night work and day sleep schedule occurs more slowly and is much less successful because the main time cue, the local daylight-darkness cycle, tends to inhibit re-synchronization. Often, re-synchronization never occurs for shiftworkers, even if they are on permanent shifts (215, 234, 278). One must wonder whether the effects of chronic jet lag on the brain may be replicated in the brains of long-time shiftworkers (52).

The Circasemidian Rhythm and the Siesta. Daily rhythms in many aspects of human performance and physiological activity actually have a two-peak daily pattern (101, 217, 237). The name of the two-peak pattern, *circasemidian* (sir ka semi dee' un), is based upon the Latin roots *circa* ("about"), *semi* ("half") and *dia* ("day"). Thus, this is a rhythm that oscillates at two cycles per day; some investigators have referred to it as the semicircadian rhythm. It usually serves to (1) deepen the pre-dawn nadir in body temperature and cognitive performance, (2) create a flat spot during the early afternoon in the daytime increase in body temperature and cognitive performance (the "post-lunch dip"), and (3) heighten the early-evening peak in body temperature and cognitive performance (202). Broughton was the first to bring this characteristic of human performance to the attention of researchers (29, 30).

No evidence exists to support the presence of a circasemidian rhythm in the rhythmic cells of the body clock, the SCN. Probably the circasemidian rhythm is the first harmonic of the circadian rhythm, generated by electro-chemical interactions among the nerve cells of the body clock and their environment. However, a number of published data sets have demonstrated the presence of the daily two-peak error pattern in industrial and transportation environments (202, 212, 216)

These behavioral and physiological observations support the biological significance of the afternoon *siesta*. Sleeping during the mid-afternoon hours, a common pre-industrial practice still followed by many cultures, prevents us from experiencing the secondary daily peak in accidents that has been observed in many industries and on the highway.

The effects of the circadian and circasemidian rhythms on accuracy and efficiency in the 24/7 workplace was obvious in the combined data from three field studies which showed that " 'real-job' speed and accuracy measures are only above average between 0700 h and 1900 h; at all other times efficiency is likely to be relatively impaired, especially so during the early hours of the morning" (109). Note the nadir at 0300 and the low spot at 1400 in the chart, below. These are the predicted effects of the combined rhythms (Figure 12).

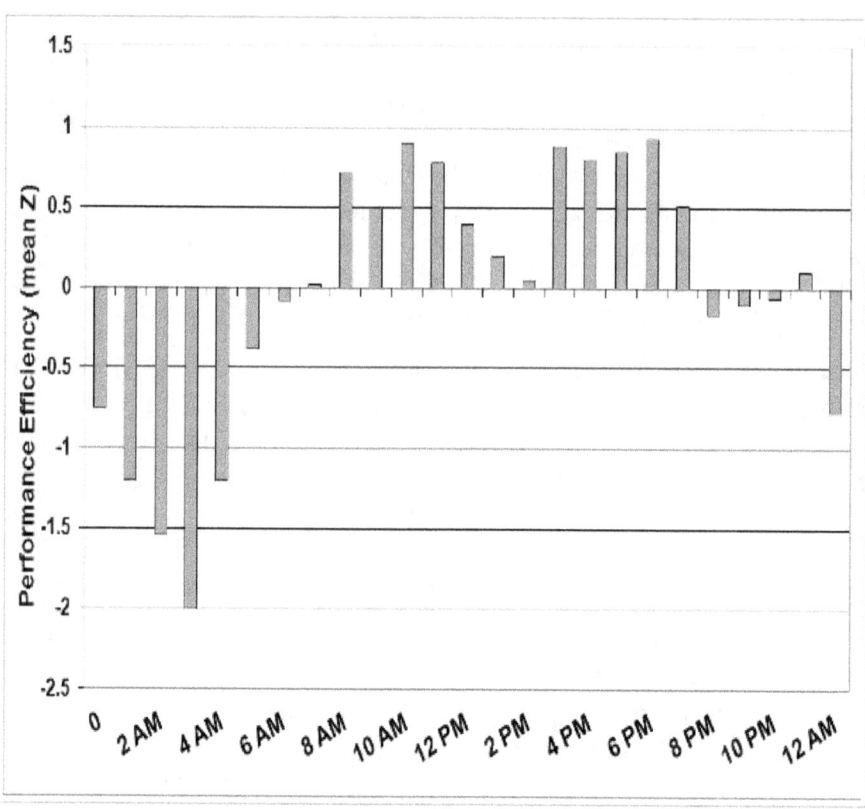

Figure 12. Relative real-job speed and accuracy measures across the hours of the day (109).

Chapter 4. Surface Watchstanding Studies 1950-1969

The earliest, relevant reports that I have found of applied research on maritime watchstanding plans come from Dr. Nathanial Kleitman. I was honored to meet Kleitman briefly in San Antonio, along with Dr. William Dement, in the early 1980s. This occurred at the opening of Dr. J. Catesby Ware's sleep clinic in San Antonio. Kleitman captured every known sleep and circadian rhythm research study in his landmark 1939 book, *Sleep and Wakefulness*, and then repeated this feat in his 1963 update of that book with 4,337 citations of research investigations (155). The latter edition was the cornerstone of my literature collection when I studied sleep and cardiovascular functions in graduate school in the early 1970s.

In starting with Kleitman's work, I hope that I have described here the beginnings of useful, modern research on the subject of the effects of watchstanding upon physiological functions and mental performance. Three observations encourage me in that hope. First, formal methods for studying human vigilance performance, a main component of watchstanding, were being developed in the period, 1948 to 1960 (179, 324).

Second, the standard method by which we define sleep stages through electroencephalography (EEG) was not published until 1968 (264). (That standard was another main component of my early-1970s literature collection. I still have the same copy!) Third, Kleitman and subsequent investigators taught research subjects to measure their own oral temperature so that their circadian rhythm in body temperature could be assessed. However, Dr. Franz Halberg, the founder of circadian rhythm research, did not publish a formal method for autorhythmometry until 1972 (121).

Thus, the methods for examining circadian rhythms, sleep, wakefulness, and mental performance were just beginning to be defined when Kleitman took his first measurements during submarine operations in the late 1940s (156).

In addition to the material presented here, I refer the reader to the excellent review produced by the Seafarer International Research Centre of Cardiff University in Wales for a wide-ranging assessment of the research literature up to the year 2000 (55). Unlike that review, I have avoided discussing the irregular hours kept by ships' pilots, focusing instead on more regular fixed and rotating watch plans.

1950

Before he became the father of sleep research in the late 1950s, and well after his famous Mammoth Cave, Kentucky, studies in 1938 of the modifiability of the human circadian rhythm, Nathaniel Kleitman of the University of Chicago was involved in several applied research projects concerning interactions between watchstanding and the circadian rhythm. At this time, Kleitman was a member of the Committee on Undersea Warfare of the National Research Council. First, Kleitman collected data during an underway study in May 1948 in the submarine, *USS Dogfish*, while the crew used the classic, fixed maritime plan (Plan 1) (156). This study is described at the beginning of my two chapters on research concerning submarine watchstanding, below.

Based upon his data from the *Dogfish*, Kleitman recommended using a "close" watch plan. The word "close" refers here to closing the gaps between watch periods so that the length of continuous time off may be expanded. In this close plan, Kleitman used work compression to allow non-watch periods of ten hours and more per section while maintaining eight hours of watch per section per day.

An underway comparison of the Close plan to a classic plan by Navy Lieutenants (jg) Utterback and Ludwig is discussed in my first chapter on submarine research, below (312). Regarding just classic plans with four hours of watch and eight hours off between watches, Peter Colquhoun and Simon Folkard summarized observations from the work of Kleitman and the subsequent underway study by Utterback and Ludwig:

> Both studies found that most individuals tended to take a sleep during *both* their 8-hour off-duty periods in each 24-hour span, since they were unable to obtain sufficient sleep during either one of them. Kleitman (1949) also observed that they rated these sleep periods as poorer in quality than those obtained ashore. The bimodality of the sleep-wake cycle was paralleled by a similar bimodality in the temperature rhythm... (63)

These observations led to Kleitman's work compression approach in his close plans. A close plan provides some longer off-duty periods in the hope that adequate sleep could be obtained in a single sleep period. In 1950, Kleitman reported investigating the ability of nine Navy recruits to adjust to various close watch plans (157). He was also interested in re-validating the observed link between body temperature and mental performance.

In this laboratory investigation, "Four watch routines were followed interchangeably in eight periods of observation, of two to four weeks each, with a total duration of five months (April through August 1949)" at the Naval Medical Research Institute in Bethesda, Maryland. Meal and sleep times varied slightly, and unrestricted coffee consumption was allowed. An "office" routine was used twice as a control condition for four weeks in May and for three weeks in August. This routine consisted of two four-hour watches per day, one from 0800 to noon, and one from 1300 to 1700. During control periods, the circadian rhythm in body temperature displayed a trough in the 0200 to 0400 period, and a peak in the mid-afternoon.

The subjects worked a dogged watch in the latter three weeks of May. Dogging (splitting) of the 1600 to 2000 watch caused watch rotation with a cycle length of four days. (Actual use of this dogged watch plan would require four watch sections.) When the subjects were off duty, there was no significant difference between this watch plan and the control for the circadian rhythm in the amplitude body temperature. The temperature rhythm was flattened somewhat when the subjects were on duty. The circadian trough times did not appear to differ from control, but the peak time drifted to about 1800. Thus, the dogged watch plan disturbed the subjects' circadian rhythms very little.

Four weeks of testing were dedicated to the classic, fixed, 4-and-8 watch plan (Plan 1). For two weeks in June the subjects worked the midnight to 0400 and noon to 1600 watches. The amplitude of the circadian rhythm in body temperature was sustained in this plan, compared to control, but the temperature trough moved later, from the 0200 to 0400 period to about 0800. This drift of the trough to a later time implies poor expected mental performance at 0800. Intuitively, we expect 0800 to be a time at which we should be starting to perform well in the morning.

For two weeks in July the subjects also worked the classic 4-and-8 plan (Plan 1), on the 0800 hours to noon and 2000 to midnight watches. The trough of the temperature rhythm persisted, compared to control, but the afternoon-evening peak disappeared. Thus, one would expect mental performance to suffer late in the day, compared to control.

In June and August, the subjects worked three close watch plans, with testing occurring within twelve-hour periods from 0800 to 2000, midnight to noon, and noon to midnight. The three close watch plans split the eight hours per day of watch period into three watches instead of two watches, and all watchstanding was completed within a twelve-hour period. This work compression allowed a twelve-hour period for recovery. Their three versions of the close watch plan might be categorized loosely as a day shift, a night shift, and a swing shift, respectively:

> Close 1: 0800 to 1100, 1300 to 1600, and 1800 to 2000; or
>
> Close 2: Midnight to 0200, 0400 to 0700, and 0900 to noon; or
>
> Close 3: Noon to 1500, 1700 to 2000, and 2200 to midnight.

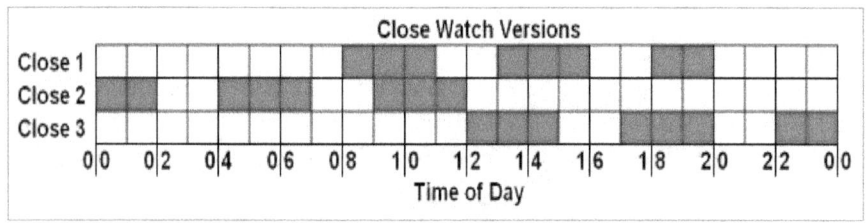

The three versions of a close watch plan used by Kleitman and Jackson (1950). Note that these are not three watch sections. Also, note that all eight hours of watchstanding is completed within a twelve-hour period in one day, allowing 12 hours off watch.

During Close 1 simulated watchstanding, the circadian rhythm in body temperature resembled the amplitude and trough time of the control group, but with a slightly delayed peak at about 2000. During Close 2, the temperature rhythm was flattened and disturbed greatly with an advanced, mid-day peak and a bizarre sharp trough in the afternoon. This was an incomplete inversion of the temperature rhythm resulting from night work. The Close 3 plan produced a temperature rhythm somewhat like the control group.

The subjects performed three mental performance tasks. The most difficult task was the maintenance of simple flight in a Link flight simulator. The subjects displayed a steep learning curve for the first month of the study, then a shallow learning curve for the remainder of the study. The subjects also performed a 15-minute alerted choice reaction time task and a five-minute color-naming task. Neither of these tasks had significant learning curves.

The Link trainer scores were remarkable with respect to the consistency of inter-subject rank order across the entire investigation. The scores were also remarkable for the very strong relationship between higher mean performance scores and lower performance variability scores. The investigators used within-subject relative performance scores for comparisons across watch plans.

The main conclusions drawn from this investigation included:

> - The recruits' temperature rhythms adjusted to new watch plans within a few days. The dogged watch required no adjustment.
>
> - Generally, the higher the body temperature, the better the mental performance.
>
> - Daily sleep amounts did not vary across the watch plans by more than half an hour per day from the eight hours per day observed in the control condition, with one exception. In one plan with afternoon leisure time, the subjects tended to sleep over ten hours per day.

As one might expect, a close plan still "suffers from the drawback that the crew whose main sleep is taken during the day may have problems in adjusting" (63). It seems that, at this point in his career, Kleitman had not seen much evidence of the recalcitrance of human circadian rhythms concerning modifiability of cycle lengths away from about 24 hours. In 1963, he wrote (155),

> From the results of these experiments it appears that there is no foundation for assuming that some cosmic forces determine the 24-hour rhythm, aside from "rest, movement, food intake, and sleep" (146). On the contrary, the rhythm seems to be *conditioned by activity of the organism which may adapt itself* to the astronomical periodicity of day and night. The revelation that some individuals offer a great resistance to the development of an artificial routine of living does not vitiate the factual demonstration that the establishment of an artificial cycle is possible. (p. 182; emphasis

added)

This opinion may have formed a portion of the basis for the decision in the 1960s to place the crews of U.S. nuclear submarines on an 18-hour "day." This topic is discussed further in the 1960s section of my first submarine chapter, below.

There seems to have been a gap in research at this point. I found no studies of watchstanding on surface ships published during the 1950s.

1962 to 1968

Robert Wilkinson and R.S. Edwards of the Medical Research Council's Applied Psychology Unit (APU) of Cambridge University in the U.K. reported conducting a twelve-day investigation of performance on a two-section plan with watch periods of seven and five hours and with major sleep in the midnight to 0700 and 1700 to midnight periods (Plan 5, below) (323, 325). They reported that "Stabilization of shift times and rotation of jobs within a shift has enabled a 2-man system of continuous manning to outperform a 3-man arrangement which lacked these features. In vigilance-type tasks calling for sustained attention it is probably the job rotation which is important; in more intense cognitive tasks such as complex decision-taking it may be the stabilization of daily shift times which confers the advantage, with job rotation doing more harm than good" (323). I have found no specification for their three-section plan, but it is likely to have been the classic maritime 4-and-8 plan (Plan 1).

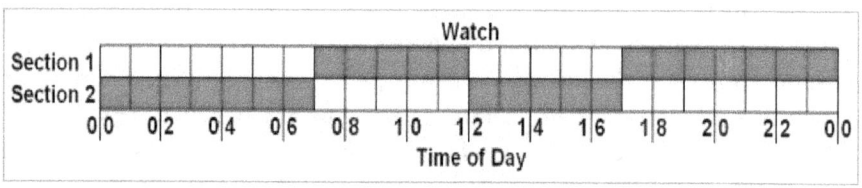

Plan 5. Two-section, fixed, 7-5-5-7 watch plan.
(Wilkinson and Edward, 1968, 1969)

My modeling of Plan 5 indicated that for section 1 the average predicted cognitive performance effectiveness across the watch periods was about 92%, with about 30% of watch time spent BCL. The body clock was stable, and one recovery night of ten hours of sleep was needed after 15 days of watchstanding. For section 2, the average cognitive performance effectiveness across the watch periods was about 70%, with all watch period spent BCL. The body clock phase advanced five hours across the

14 days and did not stabilize. Two recovery nights of ten hours of sleep were needed. Thus, as expected, section 2 fared worse than section 1.

Peter Colquhoun, Michael Blake and R.S. Edwards, also of the APU, considered four watchstanding plans in the period 1962 through 1968, reporting the results to the Royal Navy (61), and then in the journal, *Ergonomics*. The plans included a rotating three-section plan with four-hour watches, a fixed three-section plan with four-hour watches, a fixed three-section plan with eight-hour watches, and a fixed two-section plan with twelve-hour watches. For all of the studies, they assessed efficiency across a period of twelve days of watchstanding. The twelve days was a period in which the learning of a reaction-time task had appeared to be complete in a previous study by Wilkinson (324).

For the investigation of the four-hour watch length, the investigators showed two versions of the same rotating plan (Plans 6 and 7, next page). It was difficult to determine whether they used one or both. Later reports showed both plans and implied that these were traditional Royal Navy watch plans (58, 59).

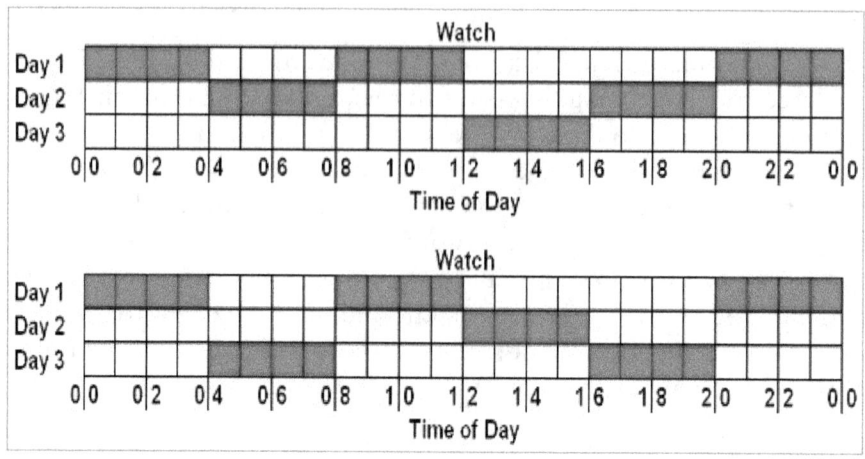

Plans 6 (upper) and 7 (lower). Three-section, rotating, 4-and-8 watch plans of Colquhoun and colleagues (61). Read each line left to right, then drop down a line for the next day. Days two and three are reversed across the two plans. In the upper plan, there is a 16-hour break (day 2 to day 3) and two four-hour breaks. In the lower plan, there are two twelve-hour breaks and two four-hour breaks. The lower plan is labeled "Rotating II" in figure 20.1 of Colquhoun and Folkard (63).

The investigators divided the four-hour watch into a vigilance task, a tea break, and a self-paced addition task. By "vigilance" in this context is meant the simple, un-alerted reaction time task, a precursor of Wilkinson's portable task (326), the precursor of today's psychomotor vigilance task, or PVT (94). Thus, the task was not a classic vigilance task that includes visual or auditory search with a low event probability. Note that these were strictly sedentary tasks. They were used, also, in the investigators' subsequent studies of eight- and twelve-hour watches.

Data were taken in the four-hour, fixed plan during 1230 to 1630 and 0000 to 0400 watches (n = 16), with sleep occurring from 0430 to 1130. In the rotating plan (n = 12), there was an obvious relationship between task performance and the circadian rhythm in body temperature. The body temperature rhythm was undisturbed in the rotating plan, but required six days to re-stabilize after the beginning of the fixed plan. In the latter case, the circadian disturbance "was accompanied by a corresponding change in the relative levels of performance" (56).

In my modeling, I noted that in Plan 6 there was one 16-hour break (day 2 to day 3) and two four-hour breaks, while in Plan 7, there were two twelve-hour breaks and two four-hour breaks. Much as in the classic three-section dogged maritime plan (Plan 4), it appeared that everyone would be fatigued. The average effectiveness for either plan was just over 80%, with about 81% and 87% of watch period spent BCL, respectively. Effectiveness declined steadily across 15 days. There was a small phase delay of the body clock, and recovery took about three nights of ten hours of sleep.

The investigators suggested that an eight-hour watch length would be preferable to a four-hour watch, and they studied task performance during a 2200 to 0600 watch (n = 16) with a mid-watch meal break. This implies the three-section plan in the figure, below (Plan 8). Sleep and a separate control watch occurred from 0800 to 1600 (n ~ 10). The body temperature rhythm flattened across the twelve-day, night-watch period.

The day-watch control provided no evidence of fatigue across the eight hours, and task performance was related to body temperature. The investigators also acquired data from a 0400-1200 watch, with planned sleep from 1930 to 0300 (n ~ 10). However, the subjects failed to sleep before about 2200 and were thus sleep deprived. However, their body temperature rhythm was not disturbed.

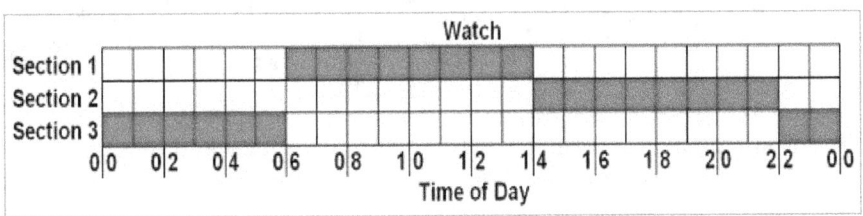

Plan 8. Three-section, fixed, 8-and-16 watch plan of Colquhoun and colleagues (60).

Section 1 in this plan was a day job (0600 to 1400 watch), but with no days off. My modeling predicted a mean effectiveness of about 96% with no watch period spent BCL. There was no predicted disturbance of the circadian rhythm, and no nights of recovery sleep were predicted. These kinds of results are the gold standard for any work plan.

Section 2 in this plan was a swing shift (1400 to 2200 watch), but again with no days off. My predicted mean effectiveness was about 96% with almost no watch period spent BCL. There was a mild disturbance predicted for the circadian rhythm, and no nights of recovery sleep were predicted to be needed.

Section 3 was the night shift (2200 to 0600 watch), which was the watch period examined in the experiment. I predicted mean effectiveness to be about 84% with about 60% of watch period spent BCL. Effectiveness declined slowly across three days and then started a very slow recovery that was nowhere near complete after 15 days of watchstanding. There was a predicted five-hour, continuous phase delay predicted for the body clock, and recovery was not complete after three nights of sleep. In the experiment, the body temperature rhythm flattened across the twelve-day, night-watch period. The model also predicted a flattened daytime rhythm of cognitive performance effectiveness during the recovery days.

Finally, the investigators examined a twelve-hour night watch from 2000 to 0800, with sleep from 1130 to 1830 (n ~ 10). This implied a two-section plan as shown, below (Plan 9). There was a separate 0800 to 2000 control watch (n ~ 10). A one-hour meal break occurred after 3.5 hours of watch, and a half-hour break occurred after eight hours of watch, resulting in 10.5 hours of actual watchstanding during the twelve-hour watch period.

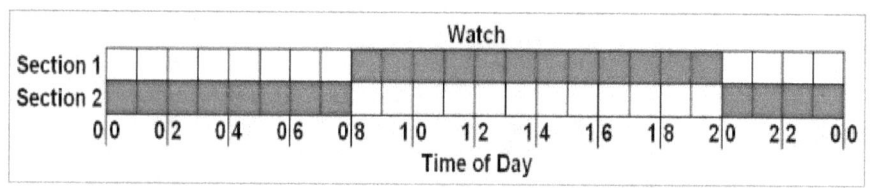

Plan 9. Two-section, fixed, 12-and-12 hour watch plan of Colquhoun and colleagues (1968).

The temperature rhythm was less flattened in the night watch than before, and intra-task vigilance performance was sustained better throughout this night watch. Again, body temperature and task performance were related. Some within-watch fatigue occurred in the day-watch control. The investigators concluded that "the relationship between temperature and efficiency was sufficiently marked to warrant further research into its generality" (62). However, looking back at studies such as these, Colquhoun (below) later expressed doubts about the validity of shore-based studies with respect to underway operations (57).

I did <u>not</u> include work breaks in modeling Plan 9, and only modeled section 2, since section 1 would have no unexpected problems with fatigue. For section 2, the average cognitive performance effectiveness across the watch periods was about 81%, with about 64% of watch period spent BCL. The body clock phase delayed six hours across 14 days and did not stabilize. More than three recovery nights of ten hours of sleep were needed.

Following their four-hour and eight-hour studies, the investigators "concluded that body-temperature was as effective a predictor of overall mental efficiency in most industrial-type shifts as in the special 4-hour shift system previously investigated" (60). While this conclusion did not speak to the issue of the best watch length, it was a seminal finding concerning the relationship between body temperature and mental performance.

Aviation investigators Dean Chiles, Earl Alluisi and Oscar Adams reported the results of an extensive research program conducted for the Aerospace Medical Research Laboratory (AMRL) at Wright-Patterson AFB in Dayton, Ohio. This program echoed maritime watchstanding research, but was aimed at the scheuling of missile silo crews (51). The work-rest plans investigated included 2-and-2, 4-and-4, 6-and-6, 8-and-8, 4-and-2, and 6-and-2. These studies lasted 96 hours each. Subsequently, these investigators collected data on the 4-and-2 plan for 15 days, on the 4-and-4 plan for up to 30 days, and they conducted other studies of these two work-rest cycles with extended sleep deprivation inserted into the plan. The investigators summarized their work as follows:

> Thirteen investigations were carried out as part of an 8-year program of research on the performance effects of various work/rest schedules during confinement to a simulated aerospace vehicle crew compartment. A total of 139 subjects were tested using a standard battery of performance tasks. ...
>
> It was found that a man can work 12 hours per day on a 4-hours work/4 hours rest schedule for periods of at least 30 days. For shorter periods, a man can work 16 hours per day on a 4/2 schedule but at a significant cost to his reserves for meeting emergencies such as sleep loss. Circadian periodicities are found in psychophysiological functions paralleled by similar periodicities in performance functions, the latter being subject to modification by special motivational instructions. (51)

The 1968 article cited here focused on the development of test methodology. In turn, the article cited the relevant government technical reports that had been published in previous years. Note that all work was sedentary. I did not model these plans because they were not associated with maritime operations.

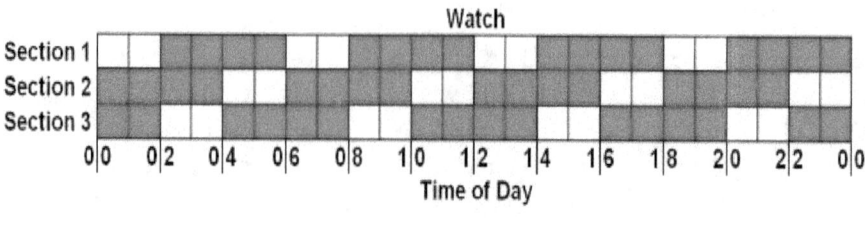

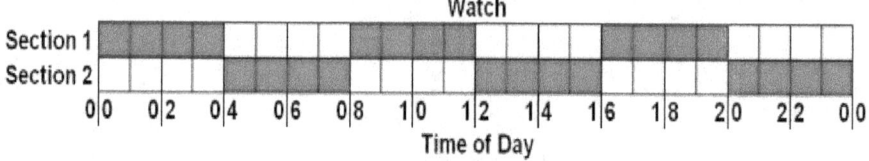

The three-section, fixed 4-and-2 (above) and two-section, fixed 4-and-4 (below) plans examined at length for the U.S. Air Force by Earl Alluisi, Dean Chiles and colleagues (3).

Chapter 5. Surface Watchstanding Studies 1970-1989

1970

Reporting at a 1970 symposium, Joseph Rutenfranz, Jürgen Aschoff (who coined the term, *Zeitgeber*), and Horace Mann, then of the Institute for Occupational Medicine at Justus Liebig University in Giessen, Germany, reported reaction time and body temperature data from twelve maritime cadets working a 4-and-8 counter-clockwise-rotating watch plan (273). I have shown this plan, below, as Plan 18, as it was also examined by Arulandam and Tsing in 2009 (discussed below).

Writing about this cadet study a few years later LaVerne Johnson and Paul Naitoh, civilian scientists with the U.S. Navy, observed (145),

> To evaluate the effects of partial sleep loss, the cadets performed a task consisting of 5 colored signals to which they responded with particular fingers, except for the fifth color to which no response was required. Each task session was brief (4 minutes long), and 40 signals were presented. The measure of task performance was mean reaction time. Oral temperature, pulse rate, and excretion of water, sodium, potassium, and calcium in urine were also evaluated.
>
> The mean sleep duration of the cadets before the voyage was 7 hours 7 minutes. During the voyage, the amount of sleep was dependent on the watch schedule. On the 2400-0400 watch, sleep was usually divided into two parts; the first period of sleep lasting 2-1/2 hours and the second approximately 2 hours, with a mean total sleep of 4 hours 52 minutes. The sleep duration on the 2000-2400 watch averaged 6 hours 4 minutes, and on the 0400-0800 watch sleep was reduced to 4 hours 38 minutes. Average sleep duration over the entire voyage was 5 hours 10 minutes, or approximately 2 hours of sleep curtailment over the 3-month period. This chronic partial sleep loss did not, however, change the task reaction time. ... The authors concluded: "the reduction in mean sleep duration to about 5 hours had no effect on mean reaction time [p.221]." (p. 23)

The phases of the circadian rhythms in mean reaction time and mean body temperature remained relatively constant with respect to midnight. Reaction times were longest at 0400, but "longer when the duration of wakefulness before the test was prolonged."

Mean reaction times were correlated negatively with mean body temperatures across the voyage, with mean body temperatures explaining 80% to 95% of the variance in mean reaction times across four measurement periods during the voyage. This is an exceptional level of agreement.

The focus here on mean reaction times in a brief test is of interest but, of course, does not address the issue of the effects of sleep deficit on higher cognitive functions, nor upon tasks of longer duration. Concerning the latter issue, Robert Wilkinson, the well-known vigilance researcher at the APU, characterized these "two quite different aspects of performance; one the ability to respond quickly and the other to sustain fast reactions over a prolonged period of work."

1976

In 1974, I had finished my doctoral course work and data collection on the combined effects of altitude and sleep on cardiovascular dynamics (203, 209, 209). As an ABD (all but dissertation) scientist, I was hired by Robert R. Mackie of Human Factors Research, Inc. (HFR) to oversee the daily operations and scientific conduct of a large-scale investigation of commercial driver fatigue (178). Subsequently, I helped conduct the U.S. Office of Naval Research (ONR) habitability study for the proposed Surface Effect Ship (SES) (239). I had two main roles in that study: implementing standardized polysomnography, as I had in my doctoral studies, and heading up the night shift for data acquisition.

At this time HFR operated the ONR Motion Generator (MoGen). The design, construction and operation of this device was overseen by HFR's chief engineer, Merlyn L. "Seltz" Seltzer (aka "The Merlin of the MoGen"). The MoGen was an 8 x 8 x 8 foot (2.4 x 2.4 x 2.4 meter) cabin mounted on a 30-foot (9.1-meter) tower. Driven by a highly-modified elevator motor system with a +/-10-ft (3-m) motion range, the MoGen was capable of generating 0 g at the top of the vertical (z) axis to 2 g at the bottom in the frequency range up to 5 Hz. The cabin also was driven in the pitch and roll axes.

The U.S. Navy sponsored extensive research concerning motion effects. To understand how motion frequency and energy induce motion sickness, see Dr. Michael McCauley's work for the U.S. Navy, conducted using the MoGen (189, 238). MoGen research showed that both lying down and sleeping are good motion sickness countermeasures; on-watch human performance is pretty much unaffected by motion sickness until vomiting begins (239).

For the SES habitability study, analog motion stimuli were created to drive the MoGen. These stimuli represented the motions expected in a 2,000-ton (2,204 metric ton) SES in three conditions: 80 knots in sea state 3, 60 knots in sea state 4, and 40 knots in sea state 5. The latter condition was especially difficult in which to try to work. We acquired cognitive performance (radar detection, encoding and decoding, navigation plotting) data as well as polysomnographic, catecholamine excretion and subjective data.

Nineteen Navy enlisted personnel entered the study, with 16 providing relatively complete data. These subjects had experienced little or no sea duty. Two subjects at a time were in motion for up to 48-hour sessions in each motion condition and also in one non-motion condition in a replicate cabin that operated in parallel time with the in-motion cabin. We had extensive problems with motion sickness that led to ad hoc modifications of the experimental design. One main issue was that when vomiting ensued as a result of motion sickness, the subjects ceased performing their tasks. Motion had no detectable effects upon cognitive performance when the subjects were working.

Each cabin was equipped with a bunk. One crewmember slept nights (2300-0030 to 0800) and the other slept days (1110-1230 to 2000). They were allowed a 1.5-hour window within which to select their bedtime. I acquired and hand-scored 21 clean polysomnographic records from 11 subjects. Motion vs. static comparisons were available for seven subjects. Statistically, there were no differences for Total Sleep Time or sleep stage percentages in motion compared to the static condition. However, motion was associated with increased stage 2. Awake time during the sleep period increased substantially in motion and with great variation at the combined expense of losses in stages 1, 3, and 4, and REM (239, Figure 14, p. 51). Thus, sleep disturbance was present in the motion condition, compared to the static condition, but was not detectable with standard statistical analyses. Four subjects experienced sleep interrupted by vomiting or severe nausea. These highly unusual occurrences are documented in the report (pp. 52-56). The circadian pattern of sleepiness ratings did not vary greatly across the static and motion conditions (239, Figure 18, p. 60). The circadian pattern of oral temperature did not differ across the static and motion conditions, but the mean oral temperature was lower in the motion condition (239, Figure 19, p. 64).

Two side notes. First, when I ran the night shift during motion operations, the attending medical officer on that shift was my good friend, the late John C. Guignard. In informal, nocturnal research John and I noted our inabilities between about 0200 and dawn to solve the 4-peg, code-breaking game *Mastermind* in our normal 4 or 5 lines of clues. We felt that this inability indicated a circadian impairment of short-term memory. This observation ties nicely into my present-day assertion that fatigue makes us stupid. But that subject is discussed elsewhere (196).

Second: eventually, the MoGen was disassembled, parked on the beach at Point Mugu Naval Air Station in California for a year, then reassembled at Michaud Naval Air Station in New Orleans. Supposedly, there was a benefit foreseen for that move. However, as usual the government's cost/benefit analysis did not consider the loss of corporate knowledge and several years of no functionality.

1978

Peter Colquhoun, P. Hamilton and R.S. Edwards at the APU reported in 1978 and subsequently an investigation of short-term watchstanding on a 4-and-4 plan (64). This was a watch plan said to have been used in some operational circumstances by the Royal Navy. It was probably Plan 5 or Plan 6, above. Twelve Navy enlisted personnel performed a continuous auditory vigilance task from 2000 to midnight and then again from 0400 to 0800 on four successive nights. The experiment was performed twice in successive weeks, with intervening rest days. On each occasion, half of the subjects slept from midnight to 0400. A control group of eleven was tested during the period 0800 to 2000. During the day, normal, acute fatigue interacted with increasing body temperature such that vigilance performance remained relatively constant across the day. At night, with declining body temperature, the "expected decline" in vigilance performance increased markedly, particularly when no sleep was allowed. Some nighttime performance decrements exceeded 50%. The investigators judged the 4-and-4 plan as not being optimal for monitoring operations for the Navy and suggested that alternative plans be considered that take into account the circadian rhythm.

1984

Peter Knauth of the University of Dortmund and colleagues, including Colquhoun, then at the U.K. Medical Research Council's Perceptual and Cognitive Performance Unit (PCPU) at the University of Sussex, and Rutenfranz, then at Dortmund, examined sleep acquired by crewmembers aboard the oil tanker *Esso Europa* while underway (158). The ship used the classic maritime watch plan (Plan 1). Twelve of the 24 crewmembers aboard volunteered for the study. Four subjects were day workers; six were watchstanders, only one of whom stood the midnight to 0400 watch; and two switched from day work to watchstanding during data collection; one of the latter stood the midnight to 0400 watch, also. The subjects provided their hours of work and sleep in diaries, recording about 175 sleep episodes.

Day workers slept at night, as expected. Those with the 12 to 4 watches acquired a major sleep period in the 0400 to noon period. Those with the 8 to 12 watches split their sleep, with a greater portion in the 2000 to 0400 period and a nap around 1400. Those with the 8 to 12 watches also split their sleep, with a greater portion in the midnight to 0800 period and also took a nap around 1400. The average durations of sleep periods were much shorter for watchstanders than for day workers. The investigators concluded, "The fixed watchkeeping system allowed only a part of the total sleep to occur between the 'normal' hours of 23.00 and 07.00, and was shown to reduce the duration of single sleep episodes."

Additionally, Ruth Condon of the PCPU and this same group reported an investigation of the circadian rhythm of oral temperature in the twelve crewmembers of the oil tanker *Esso Europa* (77). Twelve crewmembers who were watchstanders and three who were day workers provided data. Temperatures were self-measured at four-hour intervals with research quality thermometers and quality checks by the investigators. The 12 to 4 watch section (one subject) had a normal low point at 0400 and an evening peak. The 4 to 8 section had a flattened rhythm with a low point at 0400. The 8 to 12 section presented low points 0800 and midnight and a peak in the afternoon. Some, but not all of the crewmembers' circadian rhythm amplitudes were diminished compared to those expected for day work and night sleep. Thus, the phase of the circadian rhythm in body temperature seemed to acclimatize to the watchstanding schedule, but the amplitude of the circadian rhythm was depressed in more than half of the watchstanders.

Angus Craig and Ruth Condon at the PCPU had 48 subjects perform "a battery of six predominantly perceptual tasks that were relevant to bridge operations on a ship at sea" (80). Testing occurred six times between 0800 and midnight. The investigators observed, "a major part of the variation in performance can reasonably be attributed to a shift in the trade-off between speed and accuracy. ... [T]he direction of the shift is towards faster but less careful performance as the day progresses." This study is one example of an insidious effect of fatigue: the willingness to accept more risk in decision-making.

1985

Peter Colquhoun reviewed the problems associated with hours of work and watchstanding duties on ocean-going vessels sailing between distant ports (57). Research findings from submarine studies in the U.S. in 1949 and 1950 were considered, as were the studies by Colquhoun, Blake and Edwards reported in 1968 and 1969 (W. P. Colquhoun and colleagues, 1968a, 1968b, 1969a, 1969b), and the submarine watch plan comparison by Caille and Bassano. (I describe these studies, below, in the submarine research chapters.) Colquhoun discussed these findings "in relation to a proposed program of research aimed at determining the optimal system for maintaining efficiency in crews operating the modern, fully automated vessels now coming into service." This proposed program apparently resulted in a 1987 study (66) and the "Work at Sea" publications that are discussed, below.

In the same year, Colquhoun and Simon Folkard, then of the PCPU, described the status of watchstanding research (63).

> Perhaps the main finding is that stabilized systems appear to be preferable to rotating short watch systems that result in fragmented sleep and disrupted rhythms. The only advantage of these latter systems is that they avoid unfairness in the allocation of 'unsocial' hours, but it is unclear how important this is...
>
> A problem that has received little attention ... is that of individual differences ... the considerable differences in the degree to which subjects adjusted to a night watch (2200-0600) were related to the

personality of extraversion...

In sum, it would appear that little is known about the effects of the stabilized systems currently used on merchant vessels. Future studies in this area need to pay greater attention to the type of task used; where possible, they should attempt to obtain 'real job' measures of efficiency. ... The practice of flying out crew members to join ships at distant ports (see Buckley et al., 1973) also needs to be studied in view of the 'jet-lag' problem associated with it. Finally, the merits of alternative systems need to be assessed. This would require, in each case, changing not only the work-rest schedule of the mariners, but also the whole rest of their normal routine, including the timing of meals and leisure activities. Only thus would it be possible to ensure that the assessment made was a truly valid one.

1986

Condon and Colquhoun at the PCPU along with Knauth and colleagues at Dortmund examined the performance and alertness data collected during the studies reported by Knauth and colleagues, Plett and colleagues, and Fletcher and colleagues (76). Overall, performance and alertness data had been acquired from 68 subjects on seven and eight ships. The ships were using the classic fixed maritime watch plan and also used day workers. The subjects self-administered a visual-search task, a letter-cancellation task, a visual-spatial estimation task, and a subjective alertness estimation task five times daily (four times for day workers). Data were available for the 4th through 11th days of sailing.

Circadian rhythms were apparent in the response speed and alertness data. The circadian rhythms in alertness ratings at the beginnings and ends of watches did not indicate problems in phase shifting of the body clock. With no absolute scaling information presented in the report, it was difficult for me to determine whether the circadian alertness curve was flattened during watchstanding.

Accuracy on the spatial task was higher at the ends of all watch periods compared to the beginnings, especially for the 0000 to 0400 watch. Response speeds were higher, also, at the end of the 0800 to noon watch compared to the beginning of the watch. Within watch sections, response speeds were much higher for the 1600 to 2000 watch than for the 0400 to 0800 watch; and they were much lower for the midnight to 0400 watch than for the noon to 1600 watch. Interestingly, there was a more pronounced circadian rhythm for response speed at the beginning of watches than at the end.

Suspecting a speed-accuracy trade-off in the visual-spatial task, the investigators produced an "efficiency" metric by dividing speed by accuracy. (This metric surprised me a bit: generally, we combine speed and accuracy into a "throughput" metric by, inversely, dividing accuracy by speed; that is, the relative number of correct responses per unit time.) These investigators observed a rise in their efficiency metric from the beginnings to ends of watches. This observation may actually reveal a loss of throughput from the beginnings to ends of watches. Thus, the increased efficiency at the end of each watch may represent a fatigue-related willingness to accept less accuracy.

Overall, the results of the investigation indicated that the subjects' body clocks were not phase shifted from the normal day-work/night-sleep pattern by participation in the classic maritime watchstanding plan.

Anita D'Amico, Edmund Kaufman and Christine Saxe of the National Maritime Research Center, Kings Point, New York, conducted a fatigue study with the objectives "to measure the effects of sleep deprivation, time of watch, and length of time on watch on watchstander's work performance and physiological measures in a simulated open-sea watch" (86)." An additional purpose was to evaluate maritime simulation as a method for studying the effects of such variables on mariner performance." The U.S. Maritime Administration funded the study. The subjects were 25 experienced second mates who were assigned randomly to four groups:

- 7.5 hours of sleep before an 0800 to 1200 watch (n = 6),

- No nighttime sleep before an 0800 to 1200 watch (n = 6),

- 7.5 hours of sleep before a 1200 to 1600 watch (n = 6), and

- No nighttime sleep or nap before a 1200 to 1600 watch (n = 7).

The dependent measures were the speed with which the subjects sighted three traffic ships, the speed with which they detected one instrument failure, and the frequency and duration of their radar observations. The investigators also collected data on mood, sleepiness, a visual search and scan task (VSS), a digit span task, a grammatical reasoning task, and heart rate, muscle tension, physical activity, and oral temperature.

I found this section of the results quite interesting; the italics are those of the investigators:

> For all subjects combined *30% reported that they found themselves dropping off to sleep during the simulated watch, and 17% admitted that they actually slept for some length of time.* Apparently this is not uncommon in real life; more than half of the subjects (57%) said that they either tend to catch themselves dropping off or actually falling asleep while standing watch aboard their own ships.
>
> When analyzed by group, these items reveal that it was *only the no-sleep group that dropped off to sleep or fell asleep* during the experiment; none of the mates in the full-sleep group did. However, when standing on actual watch on a ship, about half of both groups reported falling asleep. (86)

Overall detection performance improved in the fourth hour of the watch compared to the first hour. However, the performance of the no-sleep afternoon group was extremely variable. Variability in performance is a hallmark of fatigue (190). Detection time for a ship astern in the middle two hours of the morning watch was much shorter than in the middle two hours of the afternoon watch. Detection time for an instrument failure in the middle two hours of the no-sleep, afternoon watch group was more than twice the time for the other three groups.

Mean responses on the Stanford Sleepiness Scale (SSS) were significantly greater for the no-sleep groups than for the sleep groups, but these means did not reach five on the seven-point SSS. Unfortunately, the investigators reported neither distributions nor non-parametric analyses of the SSS data. I suspect that there were a number of cases where individual SSS ratings were five or greater; these ratings should trigger cause for concern about safe operations.

I have not discussed all of the findings of this extensive investigation. The investigators' conclusions appropriately conservative. Here are two interesting comments from the discussion section of the report (86).

> Masters, Mates and Pilots (MMP) established union work rules requiring a deck officer to have at least six hours of rest in the past 12 hours prior to assuming a bridge watch after leaving port. According to the mates who participated in this project, this rule

> is frequently violated. They reported that it is implicit on many ships that the watch will be kept even if the mate is excessively fatigued. This policy, they say, is frequently endorsed by the master. (1978-1981 Master Collective Bargaining Agreement Covering Offshore Vessels Under Contract with the International Organization of Masters, Mates, and Pilots). (p. 79)

Also,

> One must recognize and accept that with reduced manpower on modern ships, the failure of one man to assume his watch impacts the schedule of several others. It must also be recognized that the failure of one man to keep his watch adequately may have a devastating impact on the entire ship. Severe fatigue may significantly impair the adequacy of watchkeeping. (p. 79)

These statements are virtually a premonition about the causes of the *Exxon Valdez* accident that occurred just three years later in March 1989. According to Jim Hall, then chairman of the NTSB:

> As with most accidents, we found that this catastrophic oil spill resulted from a cascade of errors and circumstances. The probable cause of the grounding of the EXXON VALDEZ was:
>
>> The failure of the third mate to properly maneuver the vessel because of fatigue and excessive workload;
>>
>> The failure of the master to provide a proper navigation watch because of impairment from alcohol;
>>
>> The failure of Exxon Shipping Company to provide a fit master and a rested and sufficient crew for the EXXON VALDEZ;
>>
>> The lack of an effective Vessel Traffic Service because of inadequate equipment and manning levels, inadequate personnel training, and deficient management oversight; and
>>
>> A lack of pilotage services
>
> (Remarks by Jim Hall, Chairman of the National Transportation Safety Board before the Washington Traffic Safety Commission Symposium on Driver Fatigue, Bend, Oregon, November 21, 1996)

Within weeks of the *Exxon Valdez* oil spill, I was aboard a tanker owned by a major international player in maritime oil shipping. We were enroute from Bellingham, Washington, to Prince William Sound. I led a privately-funded study of the fitness for duty of deck officers on the company's tankers at their request, and this was the first data collection voyage. The crew of my tanker told me the following story about the *Exxon Valdez* spill.

Mates often did not sleep much during the approximately 24 to 36 hours of ballast offloading and oil onloading that occurred at the Valdez Terminal. (The *Exxon Valdez* was at the Terminal from about 2330 until 2100 the next evening.) Upon departure fronm the terminal and early in their transit of the right-hand, southbound traffic lane through the Valdez Narrows, the access to Prince William Sound and the Terminal, they observed or were told by the local maritime traffic center of an iceberg in that lane. Captain Hazelwood asked the center for permission to continue southbound in the left-hand, northbound lane. There was no northbound traffic, so his request was granted. He observed the ship's course change to the southeast to change lanes, then went below, leaving the third mate in charge on the bridge.

The final NTSB report noted that "Accounts and interpretations differ as to events on the bridge from the time Hazelwood left his post to the moment the Exxon Valdez struck Bligh Reef. NTSB testimony by crew members and interpretations of evidence by the State of Alaska conflict in key areas, leaving the precise timing of events still a mystery." My crew reported to me that the rudder autopilot was engaged and that, when it was time to make the right-hand turn to head south and re-align with the lane, neither the mate nor the helmsman looked above the bridge windows at the rudder indicator to make sure that the rudder was answering the helm. It was not. The rudder autopilot was still engaged. By the time that the mate and the helmsman detected the problem, disengaged the rudder autopilot and initiated the right turn, it was too late, and the ship struck Bligh Reef on the east side of the northbound lane.

In the final NTSB report of the accident, it was stated that the third mate "may have been awake and generally at work for up to 18 hours preceding the accident." The accident occurred just after midnight. The third mate had been on watch for six hours, and was the only officer on the bridge. My crew was impressed by the idea that crew fatigue could lead to a disastrous failure to detect an improper configuration of the ship's controls.

1987

Led by the Karolinska Institute in Stockholm, Lars Torsvall, Kerstin Castenfors, Torbjörn Åkerstedt and Jan Fröberg reported having collected sleep-wake diaries for one to three months from 49 engineering officers in the Swedish merchant marine (303). Unlike classic watchstanding, these officers were on watch every two to four nights "during which they were allowed to sleep but were awakened by an automatic alarm system in the case of machinery malfunction." About 15% of their nights at sea were disturbed by alarms. As one might expect, watch duty, particularly with alarms, reduced their ratings of sleep length, sleep quality, and recuperation.

Colquhoun at the PCPU and colleagues K.J. Watson and D.S. Gordon collected body-temperatures and subjective alertness ratings from six subjects on a rapidly rotating watch plan (66). They compared the results to those for the rapidly rotating plan with a four-day cycle of Kleitman and Jackson (157). "The similarity of the temperature curves from the two studies indicated that prolonged exposure to shipboard conditions made no difference to the underlying circadian rhythm in this variable, which appeared to be near-normal in form." They noted that "operational effectiveness is likely to be reduced in watches held in night hours" and that "it was also likely to be reduced in daytime hours following these night watches, because of the disruption of sleep that results from them." They proposed the use of a four-section, six-hour fixed plan. See also the study by Lee and Sanquist (167), below.

1988

A series of five "Work at Sea" articles appeared in 1988 from the PCPU in the U.K. (Colquhoun, Condon) and the University of Dortmund (Knauth, Rutenfranz) dealing with circadian rhythm disruption in maritime operations using the classic 4-and-8 watch plan. Colquhoun's concern (57) about the applicability of shore-based experiments to watchkeeping underway may have been expressed in the abstract of one of those articles. "These distortions of [circadian] rhythm waveforms ... add another dimension to the basic problem caused by the effects of circadian rhythms on operational efficiency in the shipboard situation" (77). Not only were circadian rhythms affected by the irregular work and sleep patterns caused by the 4-and-8 plan, but additional shipboard demands distorted the rhythms. Shore-based experiments are not affected by these extra demands. A number of fatiguing factors other than watch plan interactions with sleep and circadian rhythms were described by Pollard and colleagues in 1990 (257, below).

In their first article, Colquhoun and colleagues provided an overview of the project and a summary of results (65). Their concerns included the effects on mental performance and ship's safety of the triad of fatigue, circadian rhythm disruptions, and sleep disruptions caused by the unusual working hours of watchstanders. They described their methodology and discussed how it was applied in a large-scale shipboard study of merchant mariners on extended voyages. They collected data from twelve of 24 members of the crew of a 250,000-ton oil tanker at sea. Nine of the twelve stood fixed watches of four hours on and eight hours off. Watches started every four hours from midnight. Meals were available from 0730 to 0830, 1130 to 1230, and 1700 to 1800. Each subject in the study provided data across eight to thirteen days. They ranged in age from late 20s to early 40s. The subjects took their own temperatures with sub-lingual thermometers, and they maintained sleep diaries provided by the investigators. The nine watchstanders had sedentary, indoor jobs.

In the second article, Rutenfranz and colleagues examined the sleep patterns of the seamen (274). They observed that "All watchkeepers exhibited fragmented sleeping patterns, which indicated a lack of adaptation of the sleep/wakefulness cycle to the hours of work. ... A solution for this problem could perhaps be a new, stabilized system that allows a single uninterrupted sleep, which is required for full recuperation, to be taken each day." In the third article, Plett and colleagues assessed the physiological variables of the seamen and concluded that, for the watchkeepers, "full phase adjustment of the circadian rhythms to shifted hours of work did not occur" (256).

In the fourth article, Condon and colleagues collected daily records of sleep, activity, body temperature, performance, and subjective alertness from 15 watchkeepers on the four-on/eight-off (1-and-3) system, and from 28 dayworkers, on both westward and eastward transatlantic voyages (77). They observed less daily sleep on the eastward voyage than on the westward voyage, and lower quality sleep. "[M]orning levels of all [circadian] variables were lower on the eastward voyage than on the westward, but evening levels were higher. ... This problem can only be solved by the development of alternative watchstanding systems which take full account of these rhythms." Finally, Fletcher and colleagues investigated the use of a compressed schedule while underway, with mixed results (108).

Temperature. One aspect of the analysis of the temperature data in the Work at Sea project was a curve fitting procedure aimed at assessing the phase and amplitude of the circadian rhythm. Franz Halberg of the University of Minnesota had developed the procedure called "cosinor" analysis (121). The investigator takes a one cycle per day cosine curve that is of maximum positive amplitude at midnight and minimum negative amplitude and noon and calculates a least-squares goodness of fit value for the temperature data. Then, by moving the peak of the curve across the 24 hours of the day and by varying its amplitude, the investigator finds the best goodness of fit value. The phase (the hours after midnight at which the peak occurs, expressed in degrees) and amplitude with the best fit is the best descriptor of the circadian rhythm of the given set of temperature data.

This method may also be repeated with cosine curves that have periods longer or shorter than one day. This produces, basically, a poor man's Fourier transform: it provides neither a complete frequency spectrum nor, directly, the relative strengths contributed by different frequencies to a complex waveform. It provides goodness of fit data for pre-selected frequency components of a waveform. Ideally, those selections have been based upon testable hypotheses.

Two caveats attend this method. First, since the circadian rhythm is not a perfect cosine curve, there is some unavoidable, unwanted, unexplained variability in the best fit. Second, as is true with many types of data reduction methods, the procedure works better on group mean data than on individual data.

Sleep. The watch plan in the Work at Sea project caused interference with sleeping, leading to the observation "that a reasonable degree of adjustment of the temperature rhythm to the different sleep/wake routines imposed by the work system had occurred." Thus, at least early in a cruise, the crew suffered from a work-schedule-induced circadian disruption (shift lag).

1989

I cite here and below several second-hand reports provided by Don Donderi, Alison Smiley and Kathy Kawaja of Human Factors North, Inc., of Toronto to Transport Canada (96). René Boulard and colleagues, as reported by Donderi and colleagues in 1995, studied crewmen during a month of Canadian Coast Guard winter search and rescue operations aboard CCGS *Simon Fraser* (26). The Boulard and colleagues report is available only in French, so I have relied upon the Donderi and colleagues description of same. According to Donderi and colleagues Canadian Coast cutters operated mainly on a two-section, fixed 6-and-6 watch plan in both 1989 and 1995. This was a two-section "layday" plan that had been introduced on CCGS *Sir Humphrey Gilbert* in 1981, with watches starting every six hours from midnight (Plan 10). There was also a 12-and-12 plan in use elsewhere.

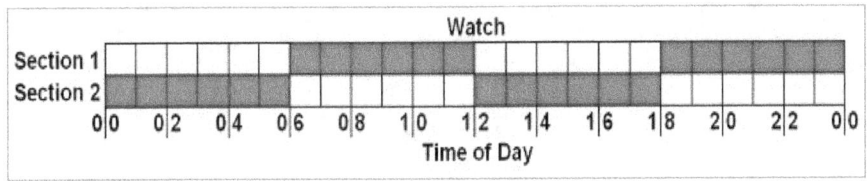

Plan 10. Two-section, fixed, 6-and-6 "layday" watch plan (26).

The small sample sizes reported by Boulard and colleagues included four crewmen on the midnight to 0600 and 1200 to 1800 watch, four crewmen on the complementary watch, and five crewmen who worked days, 0600 to 1930. The comparisons of physiology and performance among these three small groups produced no useful differences between sections. However, all crewmembers showed disturbances of the circadian rhythm, with the day workers disturbed the least.

In my modeling of section 1 of Plan 10, I found a predicted average cognitive performance effectiveness across the watch periods of about 92%, with 40% of watch period spent BCL. The body clock phase delayed one hour across 14 days and did not stabilize. Two recovery nights of ten hours of sleep were needed. For section 2, the predicted average cognitive performance effectiveness across the watch periods was about 75%, with <u>all</u> watch period spent BCL. The body clock phase delayed two hours across 14 days and did not stabilize. Three recovery nights of ten hours of sleep were needed. As expected, section 1 fared much better than section 2.

Based upon their experience with the 6-and-6 plan, Boulard and colleagues recommended the use of a 4-and-8 plan using three sections (Plan 11, below). In the <u>fixed</u>-watch plan that they recommended, each section stood a four-hour watch, then performed routine daytime work for four hours, had four hours off, stood watch again for four hours, then had eight hours off (figure, below). This is a version of the classic maritime 4-and-8 watchstanding plan (Plan 1), slipped by two hours after midnight and with designated day work and night sleep periods. The watches began every four hours from 0200 hours. The major sleep periods for the three sections occurred within the eight-hour periods, 2200 to 0600, 0200 to 10000, and 1800 to 0200, respectively.

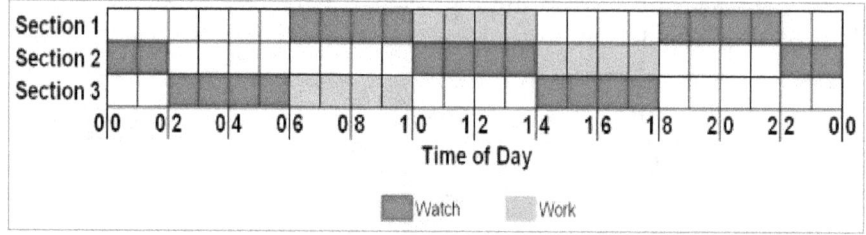

Plan 11. Three-section, fixed, 4-and-8 watch plan recommended by Boulard and colleagues (26). Watch periods are in red (dark), with the first watch for section 1 beginning at 0600. Work periods are in gray (light), with the work period for section 1 starting at 1000.

My modeling of section 1 in Plan 11 predicted an average cognitive performance effectiveness across all watches of about 95%, with only about 7% of watch periods spent BCL. There was a phase advance of about 0.5 hours over the first five days, then stability of the circadian rhythm. Recovery occurred in one night of ten hours of sleep. For section 2, the predicted average cognitive performance effectiveness across all watches was about 95%, with only about 12% of watch periods spent BCL. There was a continuing phase delay of about 5 hours across the 15 days without stability of the circadian rhythm. Recovery took about three nights of ten hours of sleep. For section 3, the 2 to 6 watch, the predicted average cognitive performance effectiveness across all watches was about 83%, with about 90% of watch periods spent BCL. There was a very small phase advance of about 0.5 hours across about 5 days, then stability of the circadian rhythm. Recovery took about one night of ten hours of sleep. These are pretty good modeling results across the three sections.

Donderi and colleagues noted a problem with the scheduling of the watch periods for senior officers and reported a resulting, operational adjustment of the Boulard and colleagues' plan. The modification was to re-arrange the eight hours of work to include six instead of four hours of daytime watch and two instead of four hours of daytime work. Of course, this implies that the second watch period for each section was reduced to two hours from four hours, and that two hours of routine work was moved elsewhere in the plan. As of 1995, this modification had not been investigated by scientists. Having inadequate information, I have not tried to reproduce the modification here. However, the Laurentian and Newfoundland extensions (below) of this modification, below, are interesting.

Boulard and colleagues also recommended a three-section, <u>rotating</u> 4-and-8 watch plan (96). This plan was based upon a five-day sequence of four-hour periods, with watches beginning every four hours from midnight (Plan 12, below). The first two days were identical, spent on the 4 to 8 watches. Then there was a four-hour phase advance to the earlier, 12 to 4 watch period on the third day, leaving only four hours of rest between watches (2000 to midnight). The fourth day was identical to the third. Then there was an eight-hour phase delay to the 8 to 12 watches on day five, leaving twelve hours between work and watch (2000 to 0800). Finally, there was a four-hour phase advance back to the 4 to 8 watches, leaving only four hours between watch periods (midnight to 0400). Note that all of day work occurred in the period 0800 to 2000.

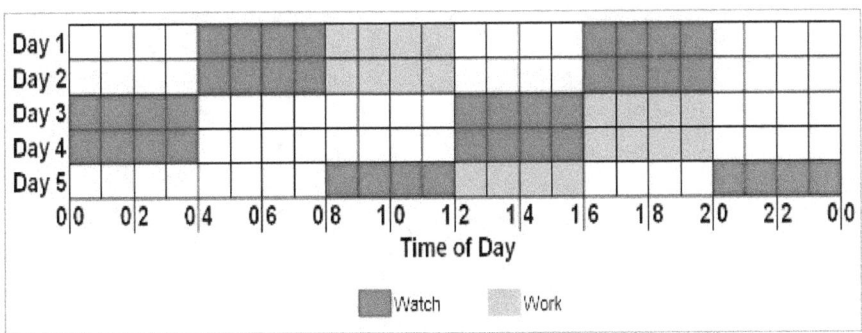

Plan 12. Three-section, rotating, 4-and-8 watch plan recommended by Boulard and colleagues for one of the sections (26). Read each line left to right, then drop down

a line for the next day.

In Plan 12, two other sections would have one day each of the 4 to 8 and 12 to 4 watches, respectively, but this section with only one day of the 8 to 12 watch would be the section with the worst plan in terms of fatigue. Alternatively, this plan may be incorporated into a 15-day cycle in which one section works all three permutations of the five-day plan. It is this latter, 15-day cycle that I modeled.

My modeling predicted an average cognitive performance effectiveness during watches of about 83%, with almost 80% of watch time spent BCL. The body clock phase delayed about three hours across the 15 days with no stabilization predicted. The latter parts of several watch periods fell below 65% effectiveness, and falling asleep on the job would probably be difficult to prevent. Recovery required about three nights of ten hours of sleep.

Chapter 6. Surface Watchstanding Studies 1990-1999

1990

John Pollard, Don Sussman and Mary Stearns at the U.S. Department of Transportation's Volpe Center examined the issue of world-wide reductions of manning on commercial vessels in an Office of Technology Assessment research report produced for the U.S. Maritime Administration (MarAd) (257). A major take-away message from this investigation, based upon interviews with about three dozen officers on five different vessels while underway, was a discussion of fatigue-inducing stressors other than the interactions of a watch plan with sleep and circadian rhythms. These stressors grouped into four main categories: organizational, voyage and scheduling, ship design, and physical and environmental.

One organizational stressor was crew continuity: "Captains and crews who had served together aboard the same ship for years had much lower levels of fatigue than those who worked on ships characterized by frequent personnel turnover." The difference here was in the amounts of supervision and training needed. Additionally, a higher level of procedural standardization was associated with lower fatigue levels. Also, as in many industries, the lure of overtime pay generated long work hours and greater fatigue levels. The reliability of ship's systems was another stressor: poor reliability demands greater amounts of maintenance and, often, less sleep.

Port calls were a critical stressor in the voyage and scheduling category: several port calls in quick succession were fatiguing due to the demands of loading and unloading. Other stressors in this category were operations in congested waters, unpredictable arrival and departure times, and long duty tours, especially those over 75 days.

The main physical and environmental stressor that caused fatigue was severe weather. This included not only heavy sea states, but also extreme heat and cold. This factor disturbed sleep and made on-deck work difficult.

In that same year, the U.S. National Research Council addressed the question, again for the Maritime Administration, of "whether smaller crews degrade safety" (227). More specifically, they asked among other questions, "Will there be greater demands placed on crew members, and if so, will they be less alert? Or, will smaller crews work the same or fewer hours and thus be no more likely to suffer fatigue than larger crews?" The Committee cited a USCG study:

> A study by the Coast Guard's Marine Investigation Division found that, between 1981 and 1985, fatigue was listed as a direct or indirect cause of casualties in only about 1 vessel in every 200 involved. However, the author noted, "It is believed that the impact of fatigue in casualties is substantially under reported as most accidents are not investigated in sufficient detail to identify its exact role." (251)

The committee found that "While concerns about safety have been raised—including neglected maintenance, increased fatigue and stress, and lessened opportunities for on-the-job training—management, labor unions, and governments have addressed these concerns through training, qualification standards, and other management techniques." The committee recommended, "The industry, with the aid of the U.S. Department of Transportation, should undertake a research program to determine how human factors such as fatigue and stress affect maritime safety." It was interesting to note that the National Research Council was recommending the kind of investigation that was already being completed by Pollard and colleagues at the Department of Transportation's Volpe Center.

Paul Naitoh, Guy Banta, Tamsin Kelly and colleagues of the U.S. Naval Health Research Center (NHRC) on Point Loma in San Diego reported an examination of the usefulness of sleep diaries to assist with the "relatively new discipline of sleep logistics (Military Sleep Management)" (224). The investigators based their sleep diary upon an original concept pioneered by Hartman and Cantrell in 1967 in my former group at Brooks AFB (201). The study indicated that sleep diaries would be quite useful for the investigation of sleep logistics during underway operations. The complementary use of wrist activity monitoring was recommended, as was the use of the "sleep fraction" estimate invented by Bill Harris and Jim O'Hanlon in 1972 at another of my former employers, Human Factors Research, Inc. (124). The sleep fraction shows the fraction of crewmembers likely to be in bed at a given time and thus not prepared for immediate action.

According to Simonia Blassingame of the U.S. Naval Postgraduate School (NPS) in Monterey, California, J.T. Lauer III in a 1991 NPS master's thesis evaluated a segment of the surface warfare officer (SWO) community assigned to shipboard command and control functions in a certain operational condition (22, 164). Lauer reported that a large percentage of the tactical action officers (TAOs) and officers of the deck (OODs) experienced symptoms of sleep deprivation while underway and a very small percentage of the TAOs and OODs felt they were fully alert at all times. The restricted distribution of Lauer's thesis prevents full publication of actual percentages; authorized persons may refer to that thesis for specific numbers (164).

1992

Karl Lumbers of the U.K. Protection and Indemnity Insurance Club (UK P&I Club) of Thomas R. Miller & Son examined patterns of major maritime insurance claims for the period 1987 to 1992 (307). In 1992, they published a circadian rhythm plot of shipping accidents. I have re-plotted their data in Figure 13.

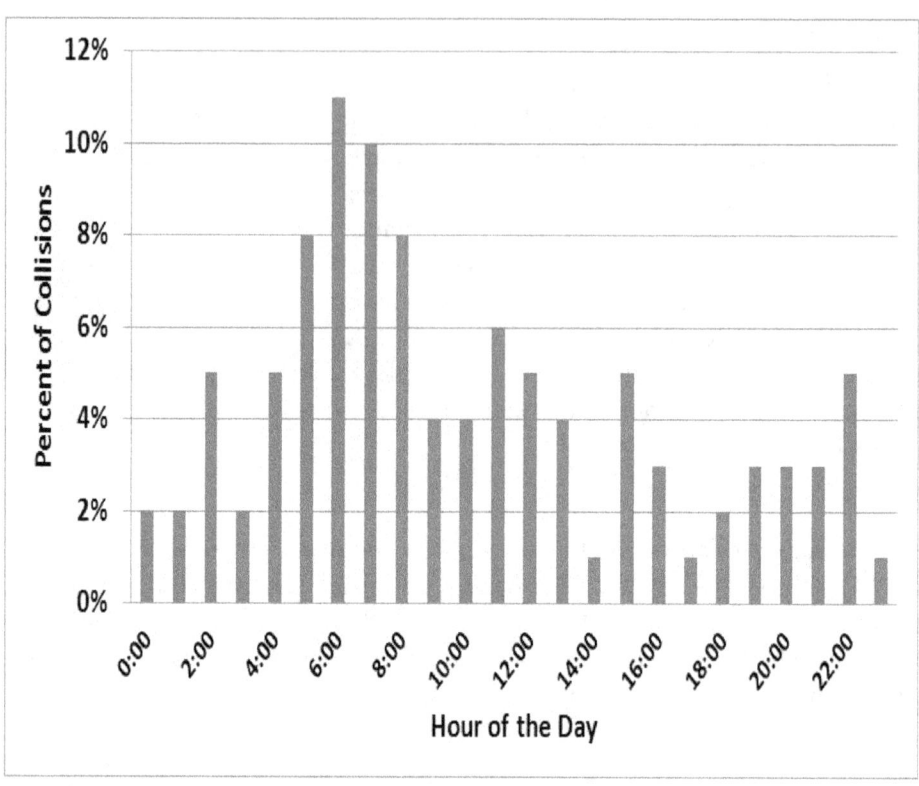

Figure 13. Percent of shipping collisions at each hour of the day (307). Note the primary peak in the early morning and a secondary peak at about 1600.

I have reproduced Lumbers' comments associated with this chart here.

> In last year's report it was stated that "second officers were involved in three times as many collision cases as other watch keeping officers". After further analysis based upon the increased database and improved information, it is now clear that there is a greater likelihood of a collision occurring during the morning watch, traditionally kept by the First Officer. This is demonstrated by [the chart, above], which shows the number of collisions, as a percentage of the total, against each hour local time.
>
> As can be seen from the [chart], 32 per cent of the collisions occurred between 0400 and 0800 with most occurring around dawn. These collisions tend to occur in areas of high shipping

activity. The Master was on the bridge in 16 out of the 21 collisions and in most cases there was a full bridge complement, including, on seven occasions, a pilot. The frequency of collisions does not yet appear to be affected by the increasing tendency for this watch to be kept by junior officers, a trend noted by the club's ship visit programme. However, it may be significant that ships frequently arrange their schedules to ensure that they arrive off port limits at dawn. (307)

Collisions are not evenly distributed through the day. Ten percent occur in one hour between 0500 hours and 0600 hours local time and altogether over 40 percent occur during the morning watch [0400 to 0800 hours]. While this may relate to the activities traditionally carried out during that watch, including entering harbor when there is a full bridge complement including a pilot, it is nevertheless worthy of further study. It does, however, reinforce the message that this watch, which frequently coincides with a change of tempo in ship's activity, is a watch where a more experienced officer might well be appropriate on occasion. (307)

I would add that it is important for the more experienced officer to be well-rested. Also, this pattern of accidents echoes many other circadian and circasemidian plots of accidents, as shown in my technical report on the circasemidian rhythm (202). There is a primary peak in the early morning hours and a secondary peak at 1500 hours. In the P&I Club data, there is another secondary peak at 2200 hours. The phase delay of the primary peak from its normal position of about 0300 or 0400 hours is suggestive of a phase delay of impaired performance associated with working through the night (41, 42). Alternatively, it may reflect an effort to avoid high-demand operations in the pre-dawn hours, postponing them until the arrival of daylight. This is an excellent fatigue countermeasure for use in operations. If it did affect the P&I Club data, then one may speculate that far more accidents would have happened during pre-dawn operations had not this precaution been taken. Part of the increase would have been due to pre-dawn fatigue effects, while part would have been due to visibility being limited by darkness.

1993

John Lee and Tom Sanquist of the Battelle National Laboratory in Seattle produced an annotated research bibliography for the U.S. Coast Guard

Research and Development Center (167). They addressed six issues: automation, crew fatigue, crew manning, organizational issues, and training.

The review "was completed for the Human Factors Planning project. It is a supporting document to the main report, 'Human Factors Plan for Maritime Safety' which will be published later." The review was divided into six sections: automation, fatigue/incapacitation, manning, navigation, organizational factors, and training. The investigators reviewed 13 fatigue studies that had been reported from the mid-1980s through 1990. I have discussed their relevant fatigue citations here previously. Regarding fatigue they stated:

> Reports of fatigue among workers in the maritime industries are common. The long work hours (i.e., greater than 8 hours per shift), the split work schedules required by the three-watch system, and environmental factors such as foul weather combine to create conditions highly conducive to acute and chronic fatigue. Perhaps the most difficult issue in determining the effects of fatigue on performance is that of definition; fatigue can be conceptualized as muscular lactic acid concentrations, perceptual efficiency, willingness to accept risk, amount of sleep loss, etc. Research on the effects of fatigue has yielded scattered results, often illustrating a dissociation between external conditions expected to result in fatigue (e.g., sleep loss) and performance effectiveness. The study of fatigue effects in the maritime industries, and the development of potential mitigation strategies Is compounded by the two classes of variable described above, i.e., sleep disruption and circadian rhythm variations. Related variables include incapacitation due to motion sickness and/or ingestion of alcohol or drugs. This section selectively reviews studies in these areas, focusing principally on fatigue.

1995

Don Donderi, Alison Smiley and Kathy Kawaja of Human Factors North in Toronto reported to Canada's Transportation Research Centre that they had found two operational "variations" on the fixed plan of Boulard and colleagues that I discussed above (96). Ships in the Laurentian area (St. Laurence Seaway) and the Newfoundland area used a day worker section, a day shift section and a night shift section (Plans 13 and 14, next page). I am hard pressed to call these two plans variations on the Boulard and colleagues plan because they both use a clever re-organization of the three sections such that one section is composed of day workers. However, the unknown originators of these plans probably adapted the "daytime ship's work" concept from Boulard and colleagues as an organizing tool. While the day workers do most of the routine ship's work, they are also scheduled to sleep through the night; this is a good thing. The Newfoundland plan is simply slipped two hours later than the Laurentian plan. Day work is constrained to the period, 0600 to 1800 or 0800 to 2000 in the two plans, respectively. The day shift and the night shift each have a six-hour watch. In the Laurentian plan, the major sleep period falls in the midnight to 0800 period for the day shift, and in the 1800 to 0200 period for the night shift. These sleep times are slipped two hours in the Newfoundland plan.

The Laurentian plan day workers would have no unexpected fatigue issues in modeling. Similarly, my modeling showed that the day watch section would probably experience about 96% cognitive performance effectiveness during their watches, with 8% of watch periods spent BCL. They would have a one-hour phase delay of the body clock over nine days, then circadian stability, with no nights of recovery required. The Laurentian night watch section would experience about 75% cognitive performance effectiveness during their watches, with all watch period spent BCL. They would be expected to experience a 2.5 hour phase delay of the body clock over about eight days, then circadian stability, with three recovery nights of ten hours sleep needed. As usual, the night watch suffers the worst fatigue effects.

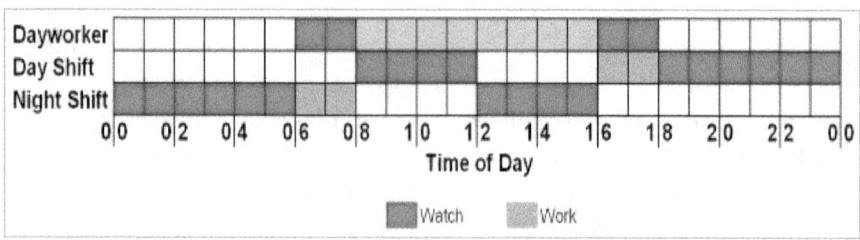

Plan 13. Three-section, fixed, Laurentian day worker watch plan (96).

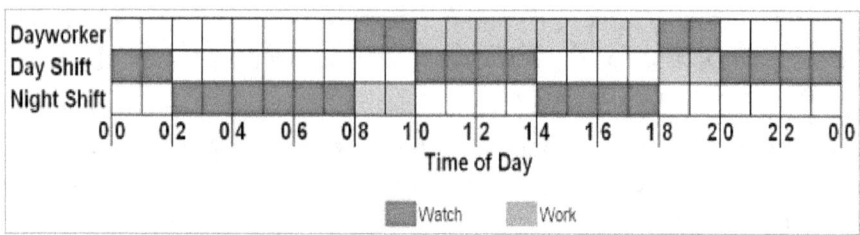

Plan 14. Three-section, fixed, Newfoundland day worker watch plan (96). This plan is simply slipped two hours later from the Laurentian plan, above.

My modeling indicated that the Newfoundland day watch section would experience about 93% average cognitive performance effectiveness during watches, with about 20% of watch time spent BCL. They would experience a three-hour phase delay of the body clock over twelve days, then circadian stability, and require three recovery nights of ten hours of sleep. This amount of predicted recovery is inordinately large for a day-work plan. The Newfoundland night watch section would experience about 81% predicted average cognitive performance effectiveness during their watches, with almost all of watch time spent BCL. They probably experienced no phase change of the body clock and required one recovery night of ten hours of sleep. This is a relatively small amount of predicted disturbance for a night watch.

In the study by Donderi and colleagues, the investigators made comparisons between the 6-and-6 layday plan and the Laurentian and Newfoundland 4-and-8 plans, above (96). They gathered data from four ships during four 28-day cruises in May through September 1994. The measures of performance were choice reaction time and short-term memory. Measures of well-being included ratings of mood, alertness and sleep quality, as well as personal sleep diaries. Small inter-group differences in choice reaction time, short-term memory, mood, sleep amounts and sleep quality favored the 6-and-6 plan. The investigators recommended a plan with day work commencing at 0800. "The two schedules [Laurentian and Newfoundland] were not significantly different in total sleep, performance (choice reaction time, short term memory) or mood. ... We do recommend that the Coast Guard consider a variation of the 4-and-8 schedule wherein the dayworker remains on the 4-and-8 schedule (which improved sleep length) and the night worker remains on the 6-and-6 schedule (6-and-6 nightworkers slept better)." Though not included the report by Donderi and colleagues, one way to do this is shown in Plan 15, below.

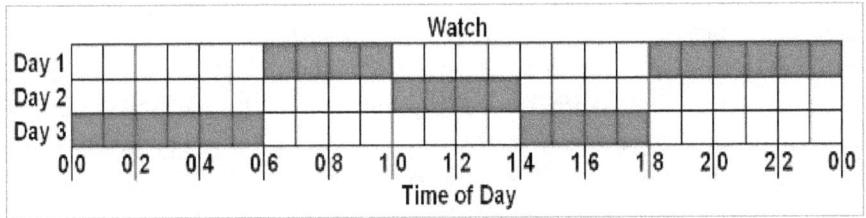

Plan 15. Three-section, combination, rotating watch plan, using 4-and-8 for days and 6-and-6 for nights (96). Read each line left to right, then drop down a line for the next day.

My modeling of Plan 15 indicated an average cognitive performance effectiveness across the watch periods of about 86%, with 71% of watch time spent BCL. The body clock would phase delay one hour across nine days, then stabilize. Three recovery nights of ten hours of sleep would be needed.

1996

Tom Sanquist, Mireille Raby and others at the Battelle Seattle Research Center and Tony Carvalhais of the U.S. Coast Guard R&D Center reported undertaking "a multi-year research program to establish a technical basis for maritime operational practice and regulatory guidance in work-rest scheduling and work hour limitations" (275, 276). The investigators noted the fatigue pattern in maritime accidents shown in the histogram, "Percent of shipping collisions at each hour of the day (307)," above. They acquired data from 141 mariners on eight commercial ships using the classic 4-and-8 fixed watch plan, above. The volunteers provided sleep and activity diary data for ten to 30 days. While the average daily sleep length at home was 7.9 hours, it was only 6.6 hours per day at sea. Watchstanders worked about 11.5 to 12 hours per day, while day workers worked about 10.8 hours per day. The data revealed three critical fatigue factors:

- 21% of sleep latencies less than five minutes [indicating extreme sleepiness]

- 8% of 24-hour periods with total sleep of less than four hours [suggesting extreme sleepiness and poor cognitive performance]

- 11% of work periods with alarmingly low alertness ratings

The investigators found that the "incidence of critical fatigue factors ... suggests that fatigue regularly occurs. The results point to sleep disruption, reduced time between watches, fragmented sleep, and long workdays as principal contributors to the problem." The investigators discussed three courses of action for fatigue reduction. The subject areas were work-rest period guidance and policy, government-industry educational programs, and design and evaluation of alternative work-rest schedules.

Marv McCallum and Mireille Raby in Seattle and Anita Rothblum of the USCG R&D Center reported an investigation of U.S. Coast Guard procedures for investigating, reporting, and analyzing fatigue contributions to marine casualties (187, 262). First, the investigators identified 279 "critical vessel" and "personnel injury" casualties with a direct human factors contribution to the casualty. These were cases in which individuals' decisions, actions, or inactions contributed _directly_ to the outcome or severity of the casualty. Second and within this data set, the investigators focused on factors contributing to fatigue and the performance consequences of fatigue. These factors included:

Mariner's experience and job position

Mariner's schedule and activities on the casualty day

A 72-hour work-rest history

Number of days off in the last 30 days

Symptoms of fatigue and contributing factors to fatigue

Mariner's decision or action

Mariner's opinion on the contribution of fatigue to the casualty

One of the clever and interesting processes used by the investigators was a factor analytic approach that helped them create a "Fatigue Index Score." The useful score had three factors:

Number of fatigue symptoms (among the symptoms: forgetful, distracted, less motivated, sore muscles, and sore eyes),

Hours worked in the last 24 hours, and

Hours slept in the last 24 hours.

Using this score, the investigators found that "Fatigue contributed to 16 percent of the critical vessel casualties and 33 percent of the personnel injury casualties."

I led an investigation of cutter crew fatigue for the U.S. Coast Guard Research and Development Center in 1996 with young Matthew Longshore Smith as my research assistant (213) (note: the USCG publication of this study does not include all of the original appendices; they are available from the author). Tom Sanquist and Marv McCallum of the Seattle laboratory and Michael McCauley of Monterey Technologies Inc. provided support and guidance as program leaders. The project provided an analysis of crew workload and fatigue on Coast Guard cutters. This was an empirical, observational study, without intervention. We acquired data during normal operations on three Reliance class (210-ft) medium endurance cutters (WMECs). These were:

USCG Alert (WMEC-630), operating along the Washington-Oregon coast; and

USCG Vigilant (WMEC-617) and

USCG Courageous (WMEC-622), both conducting operations in the Caribbean.

We supplemented our baseline analyses on the three WMECs with data from one cutter in each of two additional vessel classes:

The Hamilton class (378-ft) high endurance cutter, *USCG Midgett* (WHEC-726), conducting fishing vessel inspections in the Bering Sea during the first half of December

The Bay class (140-ft) ice-breaking tug, *USCG Katmai Bay* (WTGB-101), breaking ice at the end of the ice season on the Great Lakes and St Mary's River

The crews stood watch on the classic maritime fixed plan (Plan 1). Data collection periods on the WMEC's were 18, 19, and 18 days; 23 days on the WHEC; and 17 days on the WTGB. Sixty-two volunteers provided data on the WMECs, 23 on the WHEC and 13 on the WTGB. The volunteers' ages ranged from 22 to 40 years, with a mean of 29 years. Not all volunteers provided complete data sets.

The Quartermasters' Log–Weather Observation and Operational Summary Sheet provided us with daily and hourly weather and sea state data. A Crew Member's Daily Log provided information about the crewmembers' daily cycles of work, rest, and sleep, and ratings of fatigue, sleepiness, perceived workload, motion discomfort, and motivation. The crewmembers also wore wrist activity monitors (WAMS). These were the Actigraph brand, manufactured at that time by Precision Control Design, Inc., Ft Walton Beach, Florida, and marketed by Ambulatory Monitoring, Inc., Ardsley NY. The Actigraph used the Cole-Kripke algorithm for data reduction (53). The study's WAM methodology, especially reconciling WAM data with sleep diary reports, are available from the author.

Unfortunately, we did not acquire enough data from the crewmembers to include the WAM data in the report. However, it is probably useful to know that I conducted unpublished pre-tests of possible WAM detection of ship motion in rough sea states. My data indicated that for the waking individual there would be no problem: the default threshold and bandpass characteristics of the WAMs, which we used in that study, sensed no ship motion when the WAMs were taped to bulkheads in various locations around the ship, high, low, fore and aft. They did pick up some engine vibration when attached to thin sheet metal on a filing cabinet.

As I noted recently elsewhere, the problem with WAMs may occur for the sleeping individual. When ship motion causes untoward body motion in the bunk, then sleep disturbance could possibly be registered by a WAM. Since ship motion mainly causes the shoulders, trunk and hips to move in the bunk, it is unlikely that ship-related motions will be picked up at the wrist. On the other hand, if ship motion actually disturbs sleep such that the individual arouses and shifts body position, then such a disturbance should and will be registered by the WAM. An automated system to distinguish between the inputs of self and environmental motion in WAMs does not currently exist. One approach to solving this problem would be to acquire parallel EEG, WAM and video data for individuals sleeping in rough sea states.

We measured crewmember cognitive performance indirectly with computerized tests. We asked the crewmembers to test at least twice per day. The tests required competence in:

> Visual search mechanisms, encoding, decoding, and rote recall

> Visual pattern recognition and spatial memory, relating to crew members' abilities to use a pattern-matching approach to system failure diagnosis

> Vigilance, the ability to remain alert and watchful in a boring environment

> Visual temporal acuity, the ability to resolve rapid changes in a visual pattern

> Fine motor control and speed

Watchstanders averaged about 9.7 hours of work per day while non-watchstanders averaged about 8.3 hours per day. Overall, the work schedule caused many crewmembers to work 1.4 to 1.75 times as many hours as they would in a classic 40-hour week.

Generally, the crewmembers acquired adequate sleep with respect to their self-reported ideal amounts, but the quality of that sleep was questionable for two reasons. First, the crewmembers tended to split their sleep into more than one period per day. Watchstanders split their sleep more than non-watchstanders and received less sleep. Second, the average vigilance performance of crewmembers was impaired, suggesting a level of fatigue similar to that of laboratory subjects sleeping only five hours per night for a week.

Vigilance performance, pattern matching performance and temporal visual acuity all declined from day to day, though the crew members reported no perceptions of accumulating fatigue. This lack of perception of declining abilities mirrors a similar effect of alcohol. Most of the time, the crewmembers performed adequately on the computer-based performance tasks except in the area of vigilance. The impaired vigilance performance of the crew members was of concern as it applies to underway tasks such as the monitoring of radar, radio, engine and other systems and visual scanning by topside lookouts. Delayed or inaccurate detections in these areas can be problematic for cutter operations.

The overall, average rating of sleepiness was much closer to the description, "Losing interest in remaining awake" than to the description "Wide awake." The circadian rhythm of body temperature was somewhat suppressed in watchstanders. The crewmembers reported about the same acute changes in sleepiness across single work and watch periods as office workers. However, the watch periods were only half as long as office workdays, and the crew members worked more hours per day than office workers.

We observed evidence of mild fatigue, specifically daytime sleepiness and a degradation of vigilance performance in many crewmembers. Paragraph 4.3.5 of the report describes the ship motion effects of these underway periods on sopite syndrome (the first symptom of motions sickness) and task performance. Paragraph 4.3.6 of the report describes noise and temperature effects of these underway periods. Given high tempo operations, significant maintenance requirements, reduced crew levels, and/or sustained high sea states, we should expect a greater proportion of the crew to suffer from fatigue and its concomitant safety risks and decreased mission capability.

Following data collection and analysis, I applied principles of industrial chronohygiene to our analysis of crewmember sleep patterns, circadian rhythms, and watch plans. This analysis led me to recommend watch plan alternatives that may reduce the probability of crew daytime sleepiness and vigilance performance degradation (Appendix G of the report). The most interesting of these plans was one that I later labeled the "Close-6" plan. Discussions with crewmembers indicated that there would be no problem with six-hour watch lengths. This plan used three rotating watch sections and six-hour watch lengths in a 2W:1F system. The required 24 hours of watchstanding was compressed into four six-hour watches in two days, giving one day off in a three-day cycle (Plan 16, below). Each section works a rotating cycle of 6-on-6-off-6-on, 12-off, 6-on-6-off-6-on, 24-off. I examined this plan in a 2001 study of alternative submarine watchstanding plans (discussed in the next chapter). The Close-6 plan appeared to provide good sleep opportunities and circadian rhythm stability.

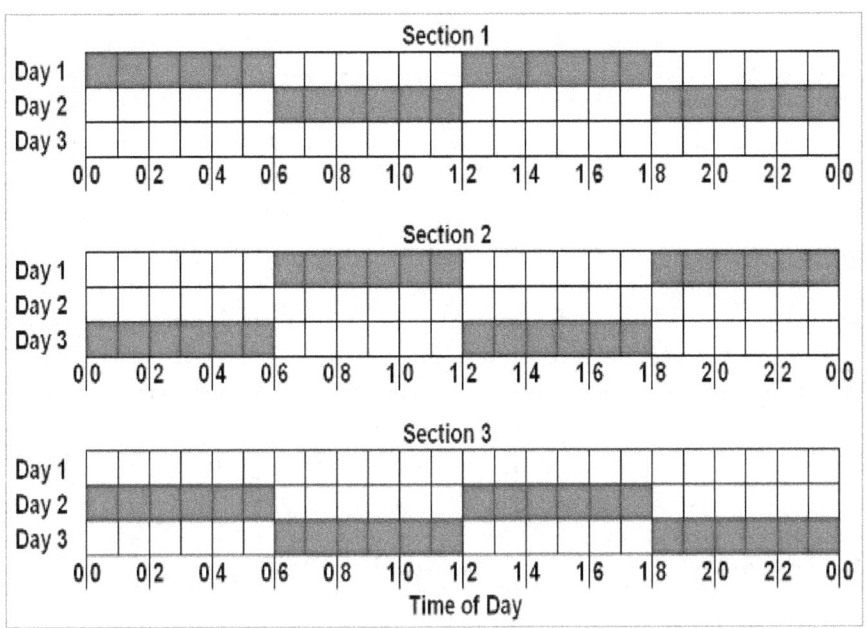

Plan 16. Three-section, rotating, Close-6 watch plan (213). In this example, the cycle starts at midnight. Alternatively, it might start at any other hour. Read each line left to right, then drop down a line for the next day.

In my Close-6 plan, one may start the plan at midnight (as in Plan 16) or at noon (213). A midnight start locates the twelve hours off from 1800 to 0600 and the 24 hours off from midnight to midnight. The latter is not optimal for recovery sleep. The noon start locates the twelve hours off from 0600 to 1800 and the 24 hours off from noon to noon. The latter period allows nighttime recovery sleep.

With a midnight start, my modeling of Plan 16 indicated a predicted average cognitive performance effectiveness across the watch periods of about 82%, with 93% of watch time spent BCL. The body clock will phase delay very slightly across six days, then stabilize. Two nights of ten hours of sleep will be needed for recovery. With a noon start, the predicted average cognitive performance effectiveness across the watch periods was about 83%, with 80% of watch time spent BCL. The body clock will phase delay 1.5 hours without stabilization by 15 days. Three nights of ten hours of sleep will be needed for recovery. Thus, the midnight start would be preferable to the noon start in terms of modeled cognitive performance effectiveness.

A personal note regarding the cutter crew fatigue study. My research assistant, Matthew Longshore Smith, and I gained a strong appreciation for the teamwork and seamanship skills of the cutter crews. We observed an excellent mix of humor and hard work in the crews. One source of humor was the fact that, invariably, I was the oldest person on board and, invariably, was assigned a top bunk in crew quarters. Another was the two of us trying to connect with a cutter in mid-patrol during a portcall, and then trying to get home after data collection. The latter included being dropped off here and there, including the northwest corner of Puerto Rico, Guantanamo Bay in Cuba, and Dutch Harbor in the Aleutian Islands. A number of our transfers to and from a cutter were by USCG helicopter. The photo, below, is of Matt and me at Cold Bay in the Aleutians the day after Thanksgiving, 1995, about to be helo'ed aboard the *Midgett* between snow squalls (Figure 14).

Figure 14. Matt Smith (left) and the author at Cold Bay, Aleutian Islands, November, 1995, awaiting helicopter transport to WHEC *Midgett*.

We had spent Thanksgiving Day in Kodiak, then flew to Cold Bay on a USCG C-130. Great fun for this old USAF C-130 pilot! The helicopter pilot from the *Midgett* impressed me, with me being a guy who used to land a 130,000-lb airplane in the dirt. The helo pilot had to dodge snow squalls and land on a pitching deck on five re-supply sorties that day. A really good pilot! [An excellent book that I happened upon in the Aleutians is Garfield's *The Thousand Mile War* (116). This is a must-read for anyone interested in the histories of WWII or the Aleutians.]

Just after this study Carlos Comperatore moved from the Army's Aviation Medical Research Laboratory at Fort Rucker, Alabama, to the USCG Research and Development Center in Groton, Connecticut. There, he conducted a couple of excellent cutter crew fatigue studies (see below, 1999) and created an excellent fatigue management program for the Coast Guard (see my chapter, below, on Fatigue Risk Management).

1997

Scott Davis, K. Hamilton, Ron Heslegrave, and Barbara Cameron of BC Research Inc., Vancouver, reported to Transport Canada an examination of crew fatigue during two 28-day (four-week) and two 42-day (six-week) cruises aboard the Canadian icebreakers *CCGS Sir Wilfrid Laurier* and *Pierre Radisson* (92). There were two twelve-hour watch sections on the *Laurier*, midnight to noon and noon to midnight. There were also day workers who worked from 0700 to 1900. On the *Radisson*, there were three watch sections, but they worked twelve hours per day with daytime overlap (Plan 16a). The *Radisson* also had day workers who worked from 0600 to 1930. For modeling, refer to the results of the modeling for Boulard's work (Plan 11).

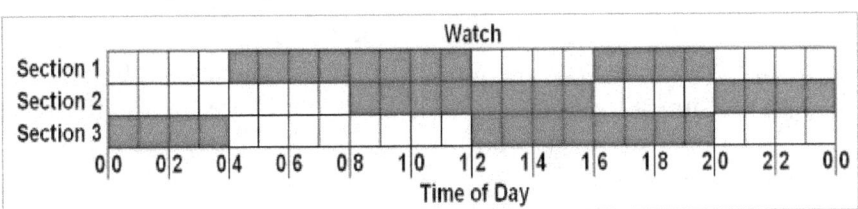

Overlapping three-section watches on the *CCGS Pierre Radisson* (92). Also, see Plan 11.

Data were acquired from 14 bridge watchstanders, 18 engineering watchstanders, and 32 day worker deck crew. The data acquired included cognitive performance (short-term memory, logical reasoning, mental arithmetic, spatial processing, and choice reaction time), mental workload, situation awareness, sleep (both diary and wrist activity monitor), fatigue, socio-psychological well-being, medications, body temperature, and symptoms experienced before and after watches (eyelid tremor).

Watchstanders experienced some perceived performance degradation, increased frustration, withdrawal, irritability, and apathy, and some reduced sleep, more so after week four. Watchstanders on the 12-and-12 plan fared slightly better than watchstanders on the 4-and-8 plan, presumably because of the occasional opportunity to get eight or more hours of sleep. The investigators noted, "Dis-entrained circadian rhythms, observed in the two case studies of the 4-and-8 watch, are a source of stress and fatigue, and ultimately performance deterioration."

The investigators extrapolated from other research showing that mental arithmetic, vigilance, reaction time, memory, and manual tracking are quite sensitive to fatigue effects. They noted, "Aspects of performance affected by fatigue which are of particular concern to marine operations are long duration tasks that are externally paced, complex, and require high levels of attention, memory and vigilance. Newly learned tasks and responses to novel situations as well as situations requiring insightful solutions are also particularly sensitive to fatigue." In their Appendix I, they applied this knowledge to the operations of the icebreakers and discussed at length the critical, specific activities required of crewmembers that were most likely to be at risk of fatigue effects.

Among the many excellent recommendations from the investigators, some interesting recommendations about watchstanding included the following:

> - The 12-and-12 plan may afford better opportunities for improved watch management through modifications to procedures during relatively low workload periods. Watchstanders should be provided with opportunities for strategic rest periods to take naps.
>
> Anticipate ship's workload so that the crew [especially watchstanders] can maximize preceding sleep and rest.
>
> Use vigilance testing if 12-and-12 watches are used across an extended cruise. [For discussions of fitness for duty testing and fatigue monitors, see (17, 117, 118, 204).]
>
> Minimize the demand for tasks sensitive to fatigue effects (per their Appendix I) whenever possible during weeks five and six.

When these tasks cannot be reduced, allocate extra time for their completion and assure that they are monitored or checked frequently.

Minimize the demand for these same tasks at night until further circadian rhythm data is collected on the 12-hour night watch.

If crew are given a day off during a cruise, encouraged them to maintain the same sleep/wake pattern as if they were working.

Assign older crew members to day watches, as aging affects an individual's ability to adapt to changes in their sleep pattern.

Provide relief watchstanders to help recovery when periods of workload are high or when conditions affect watchstanders' abilities to obtain adequate rest and sleep.

The same investigators conducted a second phase of this study (91). They collected additional data from two different crews on two cruises of the icebreaker two crews on board the icebreaker *CCGS Henry Larsen*. The Larson used two watch sections, and watches began from midnight. Data were acquired from five bridge watchstanders and five engineering watchstanders. Combining the data from the two studies, the investigators made comparisons across the following watchstanding plans:

6-and-6, *Larsen*, Phase 2

12-and-12, *Larsen*, Phase 2

4-and-8, *Radisson*, Phase 1

12-and-12 *Laurier* plus *Radisson*, Phase 1

Also, the 4-and-8 and 12-and-12 watch plans of Phase 1 were compared to the 6-and-6 and 12-and-12 watch plans of Phase 2. They also subdivided the portions of the cruises into three periods:

Days 7 to 12

Days 26 to 31

Last five days of the extended cruise

There were few significant differences across these three time intervals, and these effects could have been related to motion as well as to fatigue. Nor was it clear that one watch plan was superior to the others. However, the investigators were able to make some additional recommendations. These included:

> Implement training for sleep hardiness. [In today's terminology, this would be Cognitive Behavioral Therapy (CBT), which works quite nicely. See, for example, Dr. Gregg Jacobs' *CBT-I Conquering Insomnia* program form Harvard Medical School (CBT for Insomnia).]
>
> Base the crew work schedule on fatigue management as well as hours of work.
>
> Develop a fatigue management manual tailored to the needs of the Coast Guard. [See the chapter, below, on maritime fatigue risk management systems (FRMS).]
>
> Unless absolutely necessary, do not disrupt crew sleep for overtime duties.

Also in 1997, John Lee, Marv McCallum and colleagues at Battelle's Seattle Research Center conducted a verification and validation of a U.S. Coast Guard Crew Size Evaluation Model (CSEM), developed between 1994 and 1996 at the Battelle Center (166).

> The model [CSEM] simulates shipboard activities by specifying when each task occurs and which crew members perform it. Just as on actual ships, crew members generally perform tasks during their scheduled watch or work period, but they may also he called upon to undertake tasks during overtime periods. High priority tasks, such as docking, might even interrupt their normal sleep period. Simulating shipboard tasks produces a timeline that shows when crew members stand watch, perform maintenance, and complete any other shipboard task. The simulation output identifies the hours that crew members work each day of the simulated voyage and any instances where tasks were delayed because crew members were not available. If tasks were not performed in a timely manner or if crew members worked excessively long hours, then the crew is considered inadequate. (166)

Data collection included structured interviews with 81 crew members, observations, analysis of planned maintenance logs, and logbook data. These data were collected across two container ships and four tankers. Formal validation and sensitivity analyses led to the following findings, among others.

> The model did "provide a firm technical basis for crew size evaluation."
>
> The model was not overly sensitive to small error in task time estimates, but was sensitive to large errors in these estimates.
>
> The model worked best if crew member sleep did not occur in the model until crew members were tired, as opposed to just predicting sleep during any period of time off.
>
> The model was properly sensitive to operational variables such as increased numbers of port calls.

The investigators made several recommendations for model improvement.

1999

In a phase 3 study reported in 1999, the same BC Research investigators as above acquired data from thirteen watchstanders on a 12-and-12 plan, with watch starts at midnight and noon (90). This occurred during two more cruises with two different crews on the icebreaker *CCGS Sir Wilfrid Laurier* in the summer of 1998. The study objectives were similar to those in phase 2.

Cognitive performance was not affected significantly across the duration of the cruises. Again, it was not clear that one watch plan was better than another plan. Watchstanders on the 4-and-8 performed better on cognitive measures, watchstanders on the 6-and-6 reported better well-being, and participants on the 12-and-12 had the opportunity to obtain better sleep. As one might expect, "a different pattern of alertness between day and night watchkeepers" was observed. Thus, a recommendation was made to raise awareness of the dangers of inadequate sleep.

After an extensive review of practices and literature, Clifford Baker and Tom Malone of Carlow International and Russell Krull of Proteus Engineering presented a report to the U.S. Coast Guard regarding potential reductions in workload and staffing for deep water operations (13). Regarding crew fatigue, they noted, "Crew fatigue must be monitored and controlled. Mechanisms to monitor fatigue must be part of the ship's operational procedure, as well as specifying means to reduce fatigue once it has been identified. It is also noted here that <u>none</u> of the fleets surveyed has imposed a mechanism to formally assess crew fatigue. In all cases, *ad hoc* judgment on the part of the crew and officers was relied upon to assess fatigue." (p. 14, emphasis added)

Having read the report by Davis and colleagues (92), these authors also made the following recommendations:

> Introduction of a 12-and-12 watch plan should be considered for Arctic icebreaking operations because it offers a better opportunity for improved watch management through modifications to procedures during relatively low workload periods. The 12-and-12 watch plan also provides the opportunity for a longer, uninterrupted period to sleep.
>
> Crew preference must be considered before implementing a 12-and-12 schedule, particularly if it is in a region unfamiliar with this type of watch.
>
> If 12-and-12 watches are implemented, vigilance testing is recommended to assess the impact on human performance. (pp. C4-C5)

Carlos Comperatore, Chris Bloch and Charles Ferry at the U.S. Coast Guard R&D Center reported evaluating crew alertness and the incidence of sleep/wake cycle disruptions for crewmembers in a study called "Exemplar" aboard the WHEC *USCG Munro* during 30 days of patrol from Tokyo, Japan to Pearl Harbor, Hawaii (75). Based upon my own study of cutter crew fatigue a few years earlier, I assume that the crews stood watch on the classic maritime fixed plan (Plan 1). Forty-five volunteers participated in the study. Short-duration, electroencephalographic (EEG) alertness tests were conducted on 14 subjects (the clinical Maintenance of Wakefulness Test, or MWT) (218). Subjects maintained wakefulness for less than five minutes in 35 of 83 sessions. This is an alarming number. The five-minute criterion is used in sleep clinics as a pre-screening tool for sleep pathologies: maintenance of wakefulness for less than five minutes indicates the need for medical diagnostic testing for pathologies such as sleep apnea and narcolepsy. Sleep pathologies were not the problem on *USCG Munro*. The problem was inadequate sleep due to the watch plan.

The volunteers also wore wrist activity monitors. Eighty-three sleep-wake cycle histories were examined for volunteers with at least five consecutive workdays. The investigators created a seven-point scale, the Circadian Disruption Index (CDI; their Table 1), with which to describe the magnitude and frequency of changes in sleep and wake-up times, the duration of the sleep period, and the incidence of increased activity during work periods. In 51 of the 70 cases, the CDI was at the high end of the disruption scale: 5, 6 or 7. The investigators noted that subjects "rotating into watch schedules that disrupt the stability of the sleep/wake cycle experienced a reduction in the energy restorative value of their sleep. Rotations from daytime watch schedules into the 0000-0400 and 0400-0800 would meet such description." The MWT and CDI results shared about 50% variance.

The investigators recommended the following:

> Implement an endurance education program in the form of training on how to optimize sleep quality and prevent shift lag.
>
> Design watch schedules that minimize sleep/wake cycle disruptions.
>
> Develop a system to optimize the number of watch-qualified personnel underway to reduce the frequency of rotations into the 0000-0400 or 0400-0800 watch schedules.
>
> Implement physical improvements to sleeping areas to improve sleep quality.

Comperatore and colleagues also reported evaluating crew alertness and the incidence of sleep/wake cycle disruptions for crewmembers in a study called "Paragon" aboard the WMEC *USCG Dependable* during 32 days of low-tempo underway operations along the northeast coast of the U.S. (70). Again, based upon my own study of cutter crew fatigue a few years earlier, I assume that the crews stood watch on the classic maritime fixed plan (Plan 1, above). Twenty-five of 30 volunteers completed the study. The investigators conducted Maintenance of Wakefulness Tests on 14 subjects. Subjects maintained wakefulness for less than five minutes in 54 of 87 sessions. Again, this is an alarming number.

These volunteers wore wrist activity monitors, also. Seventy sleep-wake cycle histories were examined for volunteers with at least five consecutive workdays. In 59% of the 70 cases, the CDI was a 5, 6 or 7. The investigators noted that subjects "exposed to frequent watch rotations showed disrupted sleep associated with the 0000-0400 and 0400-0800 watch schedule." A statistical analysis showed that maintenance of wakefulness was significantly shorter when the CDI was 5, 6 or 7 than when the CDI was 1, 2 or 3 (4.7 *vs.* 9.4 minutes, respectively; $p < 0.05$). The investigators again recommended the creation of a crew endurance education program. That program is described in my chapter, below, on maritime FRMS.

Heidi Howarth, James Pratt and Don Tepas of the University of Connecticut analyzed survey data from two samples of maritime crew members (134). They found that on-duty (watchstanding) sleep length was shorter than off-duty sleep length. They also analyzed the data from a Sleep Disturbance Scale that they based upon factor-analytic scores. The correlation between the Scale score and sleep length was not significant. Thus, they suggested caution when "labeling shift workers as having 'disturbed sleep' or suffering from 'sleep disorders'."

Chapter 7. Surface Watchstanding Studies 2000-2015

2000

Richard Phillips of the University of Tasmania reported a determination of "how accident investigators report sleep in [Australian] Incident at Sea Reports and subsequently analyse the relationships between sleep, fatigue and accidents in these reports" (253). He used the Non-Numerical Unstructured Data Indexing, Searching and Theorising (NUDIST) software, a qualitative analysis program, to examine 44 incident reports. The vessels involved used the classic fixed maritime watchstanding plan (Plan 1). Reference to sleep was made in 38 of the 44 (86%) of the reports. Of these, 77% referred to incidents that occurred during the two 0000 to 0800 watches, and 49% to incidents during the 0000 to 0400 watch. Three kinds of sleep issues were noted. The worst was falling asleep on watch or being sleep deprived. The next was being awakened at odd hours, sleeping during the day, working or resting instead of sleeping, and sleep hygiene. Finally, sleeping accommodations and the sleeping routine. The investigators indicted fatigue as a contributor in eight accidents, seven of which occurred during the two watches in the 0000 to 0800 period. The predisposing factors in the fatigue-related accidents were "excessively long shifts, the nature of the work routine and disrupted sleep patterns." Sleep pathologies were mentioned. Interestingly, "jet-lag was identified as a predisposing factor where ship's officers travelled long distances to take up their command without having sufficient time to adjust to the new time zone." This is reminiscent of the recent report, for which I was a reviewer, concerning cross-country commuting of commercial airline pilots (228).

In Phillips' overview of his conclusions he stated, "Accident reports describe contributory factors such as long work periods, poor quality sleep, 'jet-lag,' broken sleep, self-inflicted sleep deprivation, disrupted sleep patterns and a poor sleeping environment. As the master, second mate and pilot were the primary contributors in two- thirds of all incidents, the routine and contribution of these particular watchkeepers could be considered in comparison to other watchkeepers." Phillips also produced an excellent flow chart of fatigue effects that may contribute to an accident (Figure 15), about which he noted:

Not all sleep is associated with the fatigue state. ... [T]he routine on board coupled with the biological need for sleep infers that all crew members probably slept from time to time during the day, hopefully, but not always while off watch. At times, ... significant crew members, such as the master ... may be asleep, when ideally their presence on watch may avert some incident. A possible relationship between sleeping, fatigue and accidents is illustrated in Fig. 1.

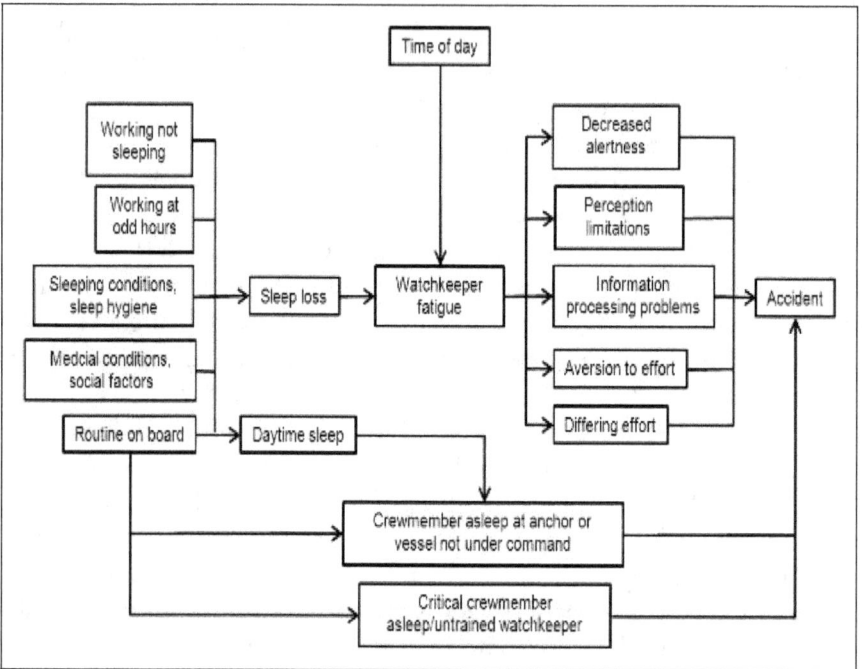

Figure 15. Sleep loss, sleeping, fatigue and accidents (re-drawn from Phillips, 2000 (253, Figure 1).

Phillips provided some other relevant observations about watchstanding:

> Watchkeeping is the responsibility of deck officers, who comprise the master and first and second mate (and perhaps a third mate), typically operating a three-watch system. The first mate takes responsibility for the 4 to 8 watch, the second mate the 12 to 4 watch and the third mate and/or the master the 8 to 12 watch. ... In addition to watchkeeping, officers have other duties; the first mate manages mooring and un-mooring operations and cargo handling, and the master has overall responsibility for command

> of the vessel and the safety of crew, specific manoeuvring operations, monitoring, budget, information control and security.
> ...
>
> Reference to these additional duties competing with sleep time is made in five instances. In one instance, the master was on the bridge for 16 out of 28.5 h but did not consider himself tired (investigators subsequently finding the master partially contributory to a collision). Another report describes a master writing a report to the shipping company at 0100 following his watch (during which the sleep-deprived pilot lost situational awareness and the vessel ran aground).

According to the Bellingham-to-Valdez oil tanker crew with whom I rode in 1989 (see above), these were primary issues for the master and third mate in the grounding of the Exxon Valdez just outside Prince William Sound, an opinion supported by the accident report (229).

Alison Collins, Victoria Matthews and Rachel McNamara of the Seafarer International Research Centre of Cardiff University in Wales published a wide-ranging assessment of the relevant research literature up to the year 2000. Some of their conclusions were as follow:

> Accident data requires closer inspection with regards hours into shift and days into tour information, although current reporting systems may not allow for this. Furthermore, information regarding exposure rates is not currently available for ... vessels, rendering prediction of accident likelihood beyond the scope of research until a uniform accident reporting procedure is put into place.
>
> The influence of environmental factors is also an under researched area, particularly amongst seafarers: the influence of poor weather and motion on performance, health and sleeping patterns can only be estimated from the existing literature.
>
> With regards performance data, the relative influence of shift timing requires attention.
>
> Examination of physiological indicators of health and fatigue have been somewhat neglected. (55)

Mirielle Raby and John Lee of the University of Iowa published an overview of maritime fatigue and workload issues (261). This book chapter provided a description of the Northwest Laboratory's 1996 merchant marine project, discussed above and conducted for the U.S. Coast Guard. The investigators concluded,

> Two field studies [166, 275] and an accident analysis [187] help reveal the causes of fatigue and its role in maritime safety. Field study data described routine ship operations and showed how these operations affect sleep and workload, and their impact on alertness and performance on board vessels. The results described here suggest that, like many 24-hour industries, fatigue poses a considerable risk to the maritime industry. The field study data also describe the task demands mat can disrupt sleep, Some of these disruptions result from what might be termed macro-level clumsy automation, where technology makes it possible to reduce crew sizes during normal operations, but may lead to workload peaks that affect the remaining crew members and can extend for several days, disrupting sleep and inducing fatigue. Evidence during normal operating conditions complements the information gathered from critical situations, which are represented in the accident analyses. The accident analysis suggested that fatigue is an important contributing factor to accidents, where it was identified as a contributing factor in 16% of vessel accidents and 33% of personnel injury cases. The field studies show that the situations that result in fatigue-related accidents are not isolated incidents in the maritime industry. Disrupted sleep and chronically low levels of alertness are characteristics of the industry that need to be carefully considered by the regulating agencies, the operating companies, and the employees that work onboard these vessels. (261)

2003

Wilfried Post, Patrick Punte, Peter Rasker and Anja Langefeld of the Netherlands Organisation for Applied Scientific Research (TNO) developed and tested an instrument for demonstrating some safety consequences of "innovative manning assignments" (259). They did not cite the previous work of Lee et al. published in 1997. The manning change to be examined by the investigators was a change to "the obligation to have a Chief Engineer on board ships with a propulsion power higher than 750 kW."

> The Chief Engineer ... cannot, for example, execute navigational tasks. Since modern ships are nowadays equipped with engines that require a minimum of maintenance and repair, the question is whether a Chief Engineer can use his time effectively on this class of ships. One possibility is to replace the Chief Engineer by a beginning maritime officer (MAROF) who has receives a dual education in navigational as well as in engineering skills and, as a result, is able to perform tasks in both areas. For example, the MAROF can take over tasks of the Master and the Chief Mate. The advantage is that this composition of the ship's crew allows a more flexible way of operating the ship. Also, labor satisfaction is expected to rise. To guarantee the original level of safety, the MAROF is backed up with 24 hours onshore technical support.

Several possible watch schedules were suggested for the changed manning structure. The primary means of data acquisition was through structured questionnaires. For this case, the investigators considered in their questionnaires:

> Conditions -- situational circumstances, personnel condition, ship condition, and shore organization
>
> Processes – navigational activities, team support, engineering activities, shore support
>
> Outcomes – safe sailing, satisfaction & atmosphere, technical functioning, realized maintenance

The result of this effort was a set of automated questionnaire instruments to be used in the subsequent twelve months. Data were to be captured 24 hours per day on 23 coasters. The follow-on publication was published in 2011 and is discussed, below.

2004

The Marine Accident Investigation Branch (MAIB) of the U.K. Department for Transport conducted an analysis of the involvement of bridge watchkeeping in 1,647 accidents that occurred from 1994 to 2003 (182). They wished to "collate the underlying human factors involved in a large number of accidents investigated by the MAIB, to graphically illustrate the principal shortfalls in bridge watchkeeping." Among many human factors questions asked by expert reviewers of the accident reports were the following:

What was person's work/rest in previous 24 and 72 hours?

Was the watch system in operation 4-and-8, or 6-and-6, or other?

What was the average voyage length/duty period?

How far into voyage/duty period before the accident?

Was the master involved in routine watchkeeping?

Was the watchkeeper sitting in a chair at the time of the accident?

Was fatigue a contributory factor

Was the watchkeeper asleep?

Was watchkeeper competency a factor?

Was the watchkeeper distracted from the main task?

Was the watchkeeper overloaded?

They found 66 accidents in which bridge watchkeeping was a factor. One-third of the 23 groundings involved a fatigued officer alone on the bridge at night. In 13 of these cases, a 6-and-6 watch plan was reported to be in use, and nine of these 13 groundings occurred in the midnight to 0500 period. In five of the groundings, a 4-and-8 plan was reported to be in use. These five groundings were spread across the period 0800 to 2000. This pattern serves to indict the two-section, 6-and-6 plan as being inadequate. There was also a general circadian pattern in the 23 groundings. That pattern was displayed in a chart that used a three-hour moving average. The chart revealed both a circadian pattern with a peak in the pre-dawn hours, and a nighttime effect, probably associated with the visibility restrictions in darkness. I have re-drawn the chart here (next page):

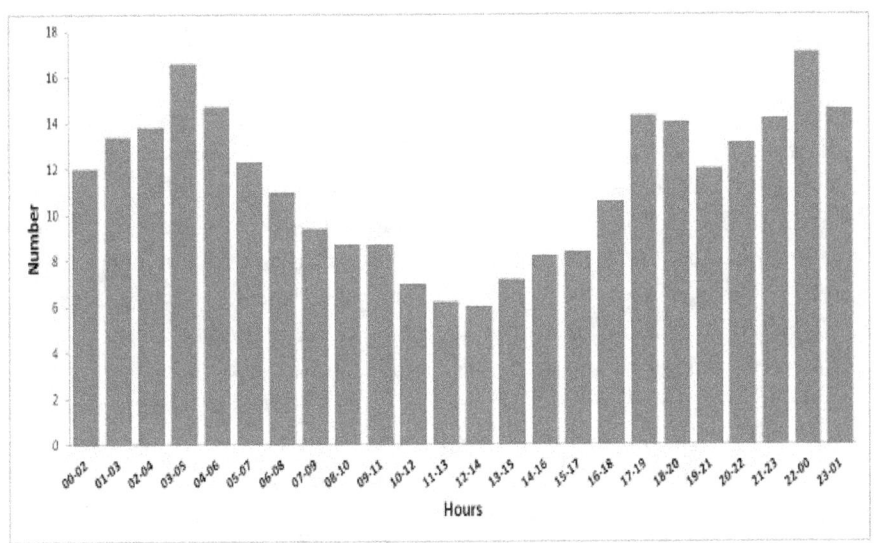

Figure 16. Circadian pattern of 23 groundings. Three-hour moving average from midnight (00-20) to midnight (23-01). Re-drawn from MAIB, 2004. (182)

Ships involved in two thirds of the 33 collisions were not keeping a proper lookout. The MAIB investigators invoked common scientific knowledge about human vigilance performance in their discussion of this problem:

> Not fully involving the lookout in the watch is detrimental to safety, and although meeting the requirement of the regulation, it does not fulfill its purpose. STCW 95 also requires that:
>
> > *The lookout must be able to give full attention to the keeping of a proper lookout and no other duties shall be undertaken or assigned which could interfere with that task.*
>
> Many OOWs [officers of the watch] seem to interpret this to mean that the lookout must stand on one side of the bridge with a pair of binoculars for between 4 and 6 hours. Performance tests [classic investigations of human vigilance performance [(211)] have shown that the alertness and concentration of lookouts diminishes after about 30 minutes, which shows the futility of employing them in this way. A proper lookout is achieved in a number of ways, not only visually; radar, AIS, radio, and telephones all need to be monitored. The provision of some

training to bridge lookouts in these areas would at least allow an OOW to use the additional manpower far more effectively.

One-third of all of the accidents that occurred at night involved a sole watchkeeper on the bridge. Most often, these were ships operating with only two bridge watchkeepers, the master and a chief officer. The investigators noted, "When at sea, working a watch on – watch off routine on the bridge, a master cannot fulfill his obligations without disruptions to his own patterns of rest, which are already disrupted by voyage cycle times." Also, "As ships operating with just two bridge watchkeepers including the master, working in opposite watches, are likely to have fatigued OOWs, and the masters of these vessels are frequently unable to discharge all of the duties required of them, the need for more than two watchkeepers is obvious."

As a result of their findings, the MAIB investigators made the following recommendations:

- [E]nsure that all merchant vessels over 500gt have a minimum of a master plus two bridge watchkeeping officers, unless specifically exempted for limited local operations as approved by the Administration.

- [C]hange the emphasis with respect to the provision of a designated lookout to ensure that a lookout is provided on the bridge at all times, unless a positive decision is taken that, in view of daylight and good visibility, low traffic density and the vessel being well clear of navigational dangers, a sole watchkeeper would be able to fulfil the task.

- Change the requirements of STCW 95 so that a bridge lookout can be used more effectively as an integral part of the bridge team.

2005

Clifford Baker and Denise McCafferty of the American Bureau of Shipping described project that was "underway to identify publicly available databases of marine accidents, review their structures, and analyze the contents. The objective of the project is to better understand the role of the human element in accident causation and consequence mitigation" (14). They used three years' of data from "accidents associated with commercial vessels in US territorial waters, as investigated by the United States Coast Guard (USCG)," plus "accident data from

Australia, Canada, Norway, and the UK." The investigators summarized their results as follow:

> While the frequency of accidents is declining, human error continues to be a dominant factor in approximately 80 to 85% of maritime accidents.
>
> Failures of situation awareness and situation assessment overwhelmingly predominate, being a causal factor in the majority of those accidents attributed to human error.
>
> Human fatigue and task omission seem closely related to failures of situation awareness and the human errors and accidents that result.

2006

In April 2006 the Swedish National Road and Transport Research Institute (VTI) published a brief, annotated research review titled "Fatigue at Sea," prepared by the World Maritime University in Malmö. One of the reviewers' conclusions was that, "With the type of research, documentation and guidance identified so far, it would seem logical that fatigue among seafarers should be a thing of the past, or at the very least, very well contained. Sadly, this is not the case." I'm not sure why they believed that fatigue would not be an issue when night work and day sleep are required of humans.

The reviewers cited and quoted recently published expressions of concern about seafarer fatigue by the UK Marine Accident Investigation Board (MAIB), NUMAST, Lloyd's List, and the UK Maritime and Coastguard Agency (MCA). The reviewers noted that "the causes of fatigue among mariners are fairly well defined and that the results of fatigue, namely accidents of varying degrees, are known. , in the maritime sector economic factors that include minimum required manning and tight schedule keeping directly affect a ship operator's ability to maintain the safest possible position regarding the fatigue of the seafarers involved." The reviewers noted several areas of concern that still required investigation.

Claire Eriksen, Mats Gillberg and Peter Vestergren of the Institute for Psychosocial Factors and Health (IPM), Stockholm, Sweden, reported an investigation of subjective sleepiness and sleep duration for a two-section 6-and-6 watch plan executed in a bridge simulator across a 66-hour experiment (106). Six officers stood the 6 to 12 watch and six officers stood the 12 to 6 watch. The two groups swapped watches halfway through the study through the interpolation of one dogged noon to 1800 watch. As one would expect, the average perception of sleepiness was significantly higher during the midnight to 0600 watch compared to the afternoon and evening watches. Sleepiness was relatively stable during the 0600 to noon watch, but increased during the other three watches. There was an initial suppression of sleepiness during all watches except 0600 to noon. This suppression may have been due to a masking effect of higher levels of activity required when taking over the responsibility of the ship. Then, sleepiness increased progressively during most watches. The investigators concluded that they were seeing initial high levels of activity within a watch period being replaced by routine activities and even by boredom. Sleep duration from 0600 to noon was 3:29 (hours:minute); from noon to 1800 it was 1:47; from 1800 to midnight it was 2:07; and from midnight to 0600 it was 4:23. These amounts align well with the influence of the circasemidian rhythm (30, 129, 202).

Josephine Arendt, Benita Middleton, Peter Williams, Gavin Francis, and Claire Luke of the University of Surrey and the British Antarctic Survey Medical Unit in the U.K. reported an investigation of sleep characteristics and circadian rhythm changes in fixed and weekly-rotating watchstanders compared to twelve day workers (7). The investigators acquired data during the underway periods of a seven-week, two-crew voyage from the United Kingdom to Antarctica aboard RRS *Ernest Shackleton*. The crews used the classic 4-and-8 watch plan with watches beginning every four hours from midnight. Six officers worked fixed shifts, while eight seamen rotated weekly to the next latest shift on each Saturday. Day workers generally worked from 0900 to 1700.

Data were acquired through sleep and activity diaries, wrist activity monitors (WAMs) and urinalysis. The latter allowed the acquisition of an indirect measure of melatonin secretion that, in turn, reflected the status of a primary human circadian rhythm. The investigators extracted four sleep parameters from the WAM and diary data: sleep duration, sleep efficiency, fragmentation index, and sleep latency. Watchstanders on the 12 to 4 and 4 to 8 watches split their sleep. The investigators labeled the two sleep periods main and secondary, depending upon sleep duration, and analyzed the two kinds of sleep periods separately.

The main sleep duration was significantly shorter in watchstanders (4.8

hours per day) than in day workers (6.0 hours per day). However, total main plus secondary sleep durations were longer for rotating watchstanders (6.7 hours per day) than for fixed watchstanders and day workers (5.9 hours per day). Sleep efficiency was poorer in the secondary sleep than in the main sleep for the watchstanders. There were few significant differences in sleep between watches in rotating watchstanders. However, rotating watchstanders displayed significantly lower sleep efficiency and higher sleep fragmentation than the other two groups. Circadian timing remained constant in day workers. Circadian timing was phase-delayed significantly for 12 to 4 watchstanders compared to the other groups and watches.

The investigators noted that none of the crewmembers had a sleep efficiency as high as those measured for good sleepers ashore. Also, the rotating plan produced worse sleep efficiency and fragmentation than the fixed plan or day work. This was probably due to the weekly circadian disruption at shift rotation. Weekly rotations have been documented elsewhere as being inappropriate for circadian stability in shiftwork. The fixed watch plan was better than the other plans in terms of sleep efficiency, fragmentation, and latency. The increase in total (main plus secondary) sleep on the rotating plan may have compensated in part for poor sleep efficiency.

The Arendt et al. data reinforced the observations of Howarth and colleagues (1999) that, though fixed watchstanders may have shorter sleep durations, the sleep quality may be better than that of day workers. However, one must note that when we are sleep deprived, we are likely to experience deeper sleep than normal. Thus, "better sleep quality" may be an indicator of a greater than usual need for recovery sleep.

The investigators of the Centre for Occupational and Health Psychology at Cardiff University have published a number of studies on fatigue at sea. Andy Smith, Paul Allen and Emma Wadsworth of the Cardiff Research Programme on Seafarers' Fatigue published an extensive report (294). That report is available free at more than 20 websites. The investigators summed up this review of the seafarer fatigue literature as follows:

> Research is increasingly revealing fatigue to be a significant problem in the seafaring industry. Present reporting systems, however, are often not designed to record this factor.
>
> Evidence shows seafarer shift and working patterns are often

conducive to fatigue. Having only two bridge watch-keepers may be a particular problem.

Excessive working hours appear widespread in the seafaring industry.

The impact of working as a seafarer may be felt in terms of health and psychosocial consequences

Research is increasingly finding a link between fatigue and shipping accidents (p. 25)

In a three-phase investigation reported here, they employed a survey that assessed health, fatigue and seafaring-specific issues, a sleep and activity diary, onboard mental performance testing (focused attention, visual search), motion and noise monitoring, sleep assessment with wrist activity monitors, and salivary cortisol assays. Their data indicated:

The results show that the potential for fatigue at sea is high due to seafarers' exposure to a large number of recognisable risk factors, both operational (e.g. port frequency), organisational (e.g. job support), and environmental (e.g. physical hazards). Our results show, however, that it is the combined effect of these risk factors that is most strongly associated with fatigue and its both short and long term consequences (fatigue symptoms, personal risk and reduced health and well-being). The most at risk groups are those exposed to the greatest number of these factors which could be identified using an audit styled approach. We have also shown that perceived fatigue is an additional risk factor for negative outcomes and this should also be included in any audit process. A taxonomic approach to fatigue should be used and measures of the frequency and intensity of different types of fatigue (e.g. acute versus chronic; physical versus mental fatigue) obtained. Appropriate tools for this have been developed and the use of measures of risk factors for fatigue and perceived fatigue will allow future associations with outcomes (e.g. accidents and injuries; health status) to be assessed. It is also important to consider personal characteristics of the seafarer to determine the extent to which these influence susceptibility to fatigue. (p. 8)

2007

In 2007 the same investigators published another excellent review article (1). One interesting aspect of this review dealt with working hours. An International Transport Federation (ITF) report had described a survey of 2,500 seafarers from 60 different nationalities. A quarter of the ITF sample reported working an average of more than 80 hours per week. A study led by Smith found 36.8% of a sample of U.K. offshore oil support ship workers worked more than 85 hours a week. The authors reported differences of opinion as to whether a goal-based or a prescriptive approach should be used to address the long work hours problem.

Concurrently, a master's student at the U.S. Navy Postgraduate School was also addressing the work hours issue. LT Leonard Haynes investigated whether the Navy Standard Workweek reflected deployed sailors' work and rest patterns accurately (130). Haynes noted,

> The 168 hours in the Navy Standard Workweek is divided into two categories: Available Time (81 hours) and Non-Available Time (87 hours). Available Time consists of tasks required to be performed by the Sailor such as standing watch and maintenance, and also includes training and attending meetings. Non-Available Time is comprised of all personal time that is allotted to the Sailor, and includes messing and sleeping.

Of the 87 hours of Non-Available Time, 56 hours per week were assumed to be time spent sleeping. During an 18-day period on the *USS Chung-Hoon* (DDG-93), Haynes collected data from 25 sailors using daily activity logs and from 22 of those sailors using wrist activity monitors (WAMs). Apparently, all of the volunteers were watchstanders, but Haynes did not specify their watch plans. At least one volunteer stood 6-and-6 watches, and Haynes used him as the typical *Chung-Hoon* volunteer; thus 6-and-6 may have been the default plan during this underway study. The sailors' reported work weeks differed quite a bit from the Standard Workweek used for Navy planning. The average reported Available Time was 94 +/- 12 hours (*vs.* 81 hours) while the average reported Non-Available Time was 74 +/- 12 hours (*vs.* 87 hours) per week.

Haynes processed the WAM data through the quantitative Fatigue Avoidance Scheduling Tool to predict the waking effectiveness level of sailors. Apparently, the Navy had used a rotating 5-and-10 ("five and dime") watch plan as a basis for the Standard Workweek. In the 5-and-10 plan, three watch sections stand four five-hour watches between 0200 and 2200 hours, and one four-hour watch from 2200 to 0200 hours (Plan 17, below). This is a good-quality plan that uses work compression (five- *vs.* four-hour watches) and clockwise rotation.

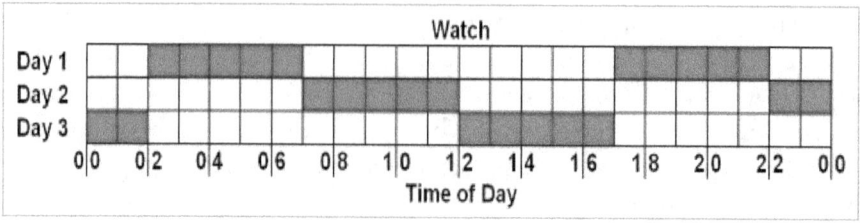

Plan 17. The three-section, rotating, 5-and-10 ("five and dime") watch plan. Three sections stand four five-hour watches between 0200 and 2200 hours, and one four-hour watch from 2200 to 0200 hours. Read each line left to right, then drop down a line for the next day.

My modeling indicated that, in the 5-and-10 watch plan, the average cognitive performance effectiveness that Haynes' predicted through FAST was slightly over 81% of good performance. My own modeling for the 5-and-10 predicted an average cognitive performance effectiveness across the watch periods of about 89%, with 51% of watch time spent BCL. The body clock remained stable, with no recovery nights needed. Alarmingly, the FAST estimate based upon Haynes' WAM data from the 22 volunteers was about 45% cognitive performance effectiveness. Apparently, the subjects' sleep quality and/or sleep quantity was much less than I entered into the model.

Haynes concluded that the Navy Standard Workweek did not reflect accurately the actual daily activities of sailors, and that the predicted cognitive performance effectiveness levels for actual watchstanding and other work were lower than the level predicted for the Navy Standard Workweek.

2008

In a third review, Allen, Wadsworth and Smith of Cardiff assessed evidence for seafarer fatigue across eleven databases, some of which were not widely available (2). They grouped fatigue-inducing factors as follow:

> Circadian rhythms
>
> Working patterns and shift schedules

Noise and motion

Sleep

Other factors

They concluded:

> Fatigue would appear to be more prevalent than the seafaring world is currently able or prepared to measure. In an industry where market competition can result in compromised standards, concern needs to be raised about pocketed crises [e.g. 18] alongside cultural malpractice undermining basic legislative safe guards [e.g. 36].
>
> Evidence suggests multiple factors are associated with fatigue at sea which is both an ecologically valid and legislatively challenging conclusion. Between shallow but exhaustive risk factor listing and single-issue campaigning the seafaring community will undoubtedly need to prioritise, implementing strategies at once both practical and policeable.

Meanwhile, Wadsworth described a data collection effort by the Cardiff group in which they used "the fatigue subscale of the profile of fatigue-related symptoms, the Cognitive Failures Questionnaire, the General Health Questionnaire and the SF36 General Health scale" (315). They received 1,855 questionnaires from seafarers working in the "offshore oil support, short-sea and deep-sea shipping industries." They found that four factors were associated consistently with both acute and longer term fatigue: "work stress, job demand, sleep quality and tour length (shorter tours)." "There was also a strong link between fatigue and poorer cognitive and health outcomes." The investigators noted that, while these findings were consistent with similar investigations of fatigue issues in demanding occupations, the study was limited by its cross-sectional design and relatively low questionnaire return rate (about 20%). They suggested longitudinal research to clarify the direction and strength of the relationships that they observed.

Phillipa Gander, Margo van den Berg and Leigh Signal of the Sleep/Wake Research Centre, Massey University, Wellington, New Zealand reported monitoring the sleep and sleepiness of 20 commercial fishermen at home and also during five to nine days at sea while these deckhands worked a nominal twelve on/six off plan (115). The investigators acquired wrist activity (WAM), sleep diary, and sleepiness rating data. The deckhands were more likely to have split sleep at sea than at home, and the median daily sleep amount at sea was 5.9 hours, compared to 6.7 hours at home. Even on the 12-and-6 plan, they preferred to sleep at night at sea. During about one-quarter of all days at sea, the deckhands slept less than four hours per day. Sleepiness ratings improved significantly less after sleep at sea than after sleep at home. The investigators concluded that "impairment due to acute and cumulative sleep restriction would have been relatively common" while at sea. A 12-and-6 work-rest plan implies an 18-hour day and concomitant, strong disturbance of the circadian rhythm. This is a problem that is addressed in the chapter on submarine crew watchstanding, below.

As a follow-on to three previous reviews (31, 55, 114), Paul Allen and colleagues at Cardiff searched for fatigue-related research since the year 2000 in eleven databases (2). Allen and colleagues noted "a useful working definition is included in the International Maritime Organization's (IMO) guidelines on fatigue [4] as follows: 'A reduction in physical and/or mental capability as the result of physical, mental or emotional exertion which may impair nearly all physical abilities including: strength; speed; reaction time; coordination; decision making; or balance.'" The authors also noted, in continuing agreement with Brown and with Collins and colleagues:

> Fatigue is only likely to be tackled as a serious issue in the maritime industry once a reliable picture concerning its prevalence is established. Unfortunately, however, a reliable picture concerning prevalence can only be built up once fatigue is taken seriously enough to warrant accurate and reliable reporting systems. Such a state of affairs has meant that research into fatigue prevalence has often been localised and vessel specific.

Also:

> Considering seafarers as an homogenous population is clearly inappropriate where working patterns and way of life can vary enormously according to a number of factors including cargo, type of trade, crew nationality and flag of registration. (2)

Mikko Härmä, Markku Partinen, Risto Repo, Matti Sorsa, and Pertti Siivonen of Finland reported examining fatigue and sleepiness in Finnish bridge officers, 42% of whom worked two four-hour watches per day, while 26% worked two six-hour watches per day (122). Both watch plans began at midnight with a midnight to 0400 watch or a midnight to 0600 watch, respectively. Ninety-two officers completed a sleep diary for seven consecutive days at sea. Their ratings on the Karolinska Sleepiness Scale (KSS) followed a very nice circadian pattern, peaking at 0400, and with a hint of a circasemidian peak at 1800 (investigators' figure 1). Both the circadian pattern and the group average KSS ratings within watches ranged from only 2 to 5 on the 9-point scale, with 1 being "very alert" and 5 being "not sleepy nor alert." The average ratings of about 5 occurred at the ends of evening and night watches and the beginnings of night watches (investigators' figure 2).

Pre-watch main sleep durations preceding six-hour watches ranged from 197 min (3:17) before the 1800 to midnight watch, to 279 min (4:39) before the 0600 watch. Preceding four-hour watches, they ranged from 224 min (3:44) before the 2000 watch to 335 min (5:35) before the noon watch. In addition to the officers working the 6-and-6 plan reporting shorter main sleep durations compared to the 4-and-8 watch plan, they also reported more frequent nodding-off on duty and more excessive sleepiness. They were able to conclude, "The results suggest the 6-and-6 watch system is related to a higher risk of severe sleepiness during the early morning hours compared to the 4-and-8."

2009

LT Derek Mason, a master's student at the U.S. Navy Postgraduate School (NPS), compared the Navy Standard Workweek to the work-rest patterns of sailors on two U.S. Navy cruisers (185). This was an extension of the investigation by Haynes (2007) conducted on a Navy destroyer, but on a larger vessel. Thirty-nine sailors onboard *USS Lake Erie* (CG-70) and *USS Port Royal* (CG-73) wore wrist activity monitors (WAMs) for 24 days and completed sleep and activity diaries. Mason analyzed the WAM data with the Fatigue Avoidance Scheduling Tool (FAST). Two of Mason's more fascinating findings were the numbers of hours slept per day as a function of rank and as a function of department (Mason's figures 8 and 9). I have reproduced his striking bar charts here (Figure 17).

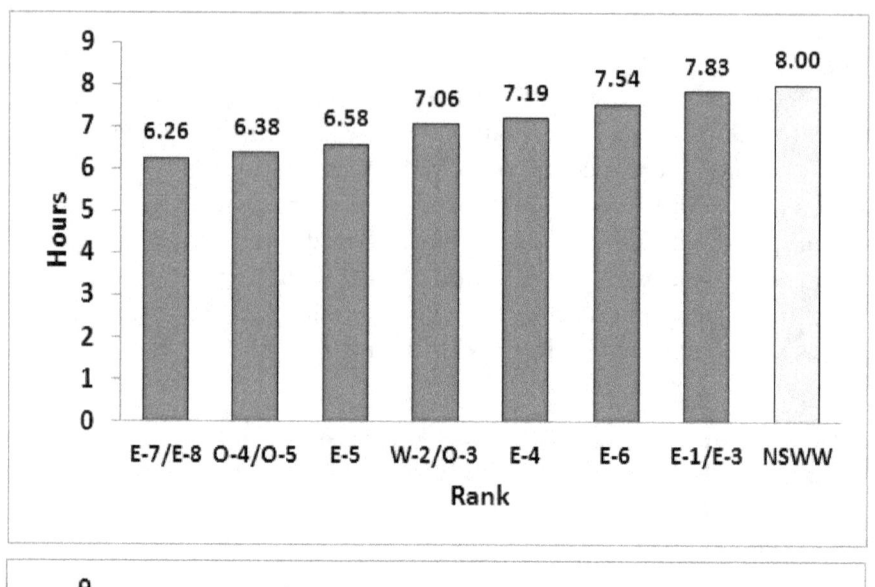

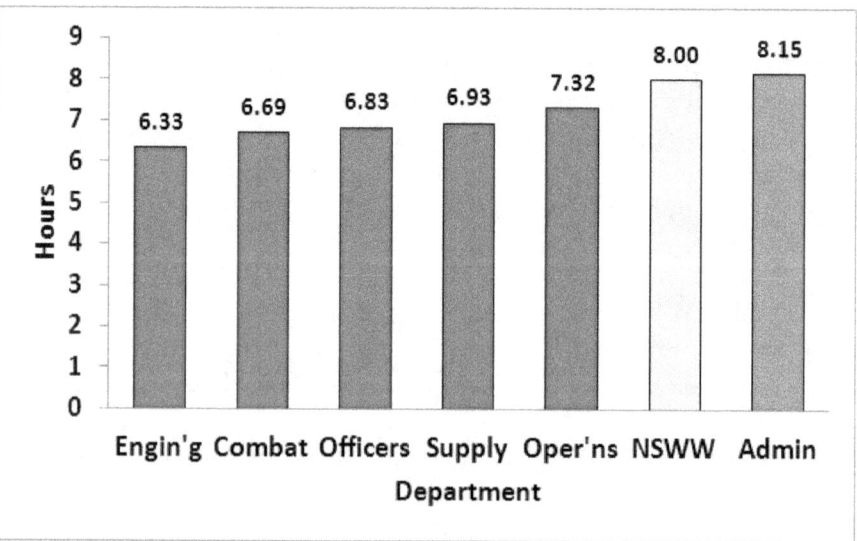

Figure 17. Numbers of hours slept per day as a function of rank and as a function of department (re-drawn from Mason, 2009, figures 8 and 9) (185). NSWW = Navy Standard Workweek.

The senior enlisted personnel (E-7 and E-8) slept the least, at 6.26 hours per day, with the senior officers (O-4 and O-5) getting little more sleep at 6.38 hours. The warrant and more junior officers (W-2 though O-3) were in the middle of the distribution at 7.06 hours. The lowest ranks (E-1 through E-3) acquired the highest amount of daily sleep at 7.83 hours. No rank group achieved the target of eight hours of sleep per day.

The watchstanders in the Engineering and Combat Systems departments acquired the least amounts of daily sleep at 6.33 and 6.69 hours. The Operations and Administrative department sailors acquired the most sleep at 7.32 and 8.15 hours per day. All but one department fell short of the eight hours per day target. Mason did not specify the watchstanding plan. Operations personnel stood watch an average of 9.15 hours per day, while personnel in other departments stood watch an average of about eight hours per day.

Mason provided this summary. "85% of the participants within the study exceeded the 81 hours of available time allotted by the Standard Navy Workweek. On average, Sailors in the current study, excluding officers, worked 9.90 hours per week more than allotted in the Navy Standard Workweek."

In another NPS masters thesis, Kim Green expanded upon the Navy Standard Workweek (NSWW) studies of Haynes (2007) and Mason (2009) with an examination of sailors' self-reported activities onboard U.S. Navy destroyers and cruisers (120). Green's objective was to determine whether similar patterns would exist onboard U.S. Navy frigates. Green's methods were similar to those used by Haynes and Mason. Three weeks of data were acquired from 24 sailors while underway on the *USS Rentz* (FFG 46). The Administration and Combat Systems Departments required about 78 hours per week of Available Time, slightly under the 81 hours allotted in the Standard Workweek, while the Supply Department required about 82 hours. The Operations and Engineering departments required 99 and 104 hours, respectively. "Results indicated that 61% of the participants exceeded the 81 hours of Available Time (work) allotted by the NSWW." Enlisted sailors worked an average of 20 hours more per week than allotted in the Standard Workweek and slept 9 fewer hours per week than allotted.

Shalini Arulanandam and Gregory Chan Chung Tsing of Singapore conducted a pilot study for the Singaporan Navy to compare alertness levels in 24 crewmembers working a fixed watch plan and then a rotating watch plan (8). For four days the volunteers followed a standard three-section maritime plan of four-hour watches beginning at midnight (probably Plan 1, above). Then the volunteers spent six days without watchstanding and, finally, four days on an unusual three-section rotating watch plan. The latter plan rotated counterclockwise by virtue of a dogging of the 1600 to 2000 hours watch (Plan 18, below).

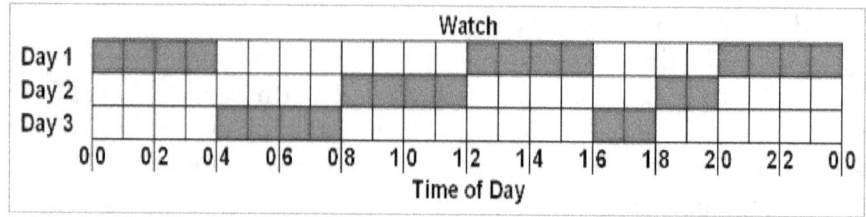

Plan 18. Three-section, counterclockwise-dogged, 4-and-8 watch plan of Arulanandam and Tsing (8) and of Rutenfranz and colleagues (273). Read each line left to right, then drop down a line for the next day.

The investigators assessed alertness levels before and after each duty watch using measurements of oculomotor function provided by the Fitness Impairment Tester (FIT-2000) (208, 272). The investigators noted, "Saccadic velocity was shown to have the greatest correlation with duration of sleep deprivation and was significantly slower (indicating decreased alertness) in the crew working the rotating watch plan than the crew working the fixed watch plan." They concluded, "This pilot study corroborates previous studies' recommendations that fixed watch plans allow better acclimatization of sleep patterns, thus minimizing fatigue and increasing operational alertness." I must caveat their conclusion by noting that the counterclockwise rotation in the plan probably disturbed the volunteers' circadian rhythms greatly. My modeling indicated a predicted average cognitive performance effectiveness across the watch periods of about 82%, with 85% of watch time spent BCL. The body clock phase delayed one hour across five days and then stabilized. Two recovery nights of ten hours of sleep were needed.

2010

Captain Nick Nash described his approach to a watchstanding plan (Plan 19) for a cruise ship (226). Having considered a number of possibilities,

> I favoured the Swedish 6,6,4,4,4. This allows a three-day cycle which gives a maximum work period of six hours – allows periods of rest (that is to say, not on watch) of 12, 10 and eight hours.
>
> I particularly like the two six-hour watches to cover the day

(08:00-20:00) which allows for the night watches to be of four hours' duration and aligned with the traditional night watches pattern (although with different officer teams). It also allows the time zone changes to be split equally between each night watch in the usual 20 minute pattern.

I planned the change of watch at 08:00 in the morning as on this cruise ship most of our arrivals are on the four to eight watch. I had previously tried an 06:00 shift change to allow the two six-hour periods to be 06:00-18:00; however I found the change at 02:00 and 06:00 unpopular with the team and in addition, the 06:00 watch change upset arrivals.

Watches were therefore set up as follows:

    04:00-08:00 (4): (Followed by 12 hours' rest)

    08:00-14:00 (6): (Followed by 10 hours' rest)

    14:00-20:00 (6): (Followed by eight hours' rest)

    20:00-24:00 (4): (Followed by eight hours' rest)

    24:00-04:00 (4): (Followed by 10 hours' rest) (226)

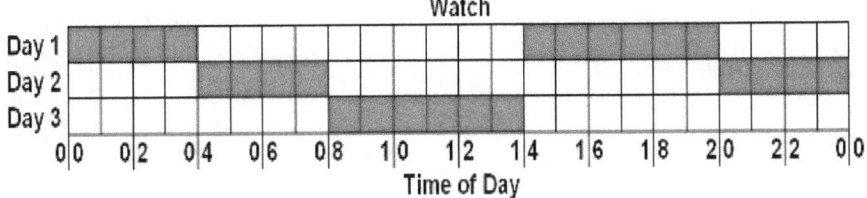

Plan 19. Three-section, 4-6-4-4-6, rotating cruise ship watch plan (226).

Nash and the ship's doctor, Mark Mason, created a questionnaire composed of 30 questions that addressed four areas: work and ILO, rest and sleep, general wellbeing, and overall satisfaction with this alternative plan. Six officers provided responses on a five-point Likert-like scale of *strongly disagree* (minus 2) through *indifferent* (0) to *strongly agree* (plus 2). The mean result for each of the four areas fell between +0.6 and +1 (*agree*). Thus, the plan appeared to be moderately well accepted.

In my modeling, I allowed a two hour nap, from 1400 to 1600, after the 0400 to 0800 watch. My results indicated a predicted average cognitive performance effectiveness across the watch periods of about 88%, with 52% of watch time spent BCL. Predicted cognitive performance during the 0000 to 0400 watch was just as terrible as usual. The body clock phase delayed two hours across 13 days. No recovery nights were needed. Pretty good results, all considered. Additionally, this is quite a good article regarding the interaction between watch planning and the command and control of a ship. More about this interaction in the chapter on Maritime FRMS, below.

Margareta Lützhöft, Anna Dahlgren, Albert Kircher, Birgitta Thorslund, and Mats Gillberg of Chalmers University in Gothenberg noted that "many merchant ships sail with only two nautical officers, working a shift schedule of 6 hr on and 6 hr off" (175)(abstract only). Across 13 ships, they collected data from 15 volunteers working a 6-and-6 watch plan and another 15 working a 4-and-8 watch plan. They used electrooculography, wrist activity monitors, sleep and activity diaries, and reaction time test. They observed higher sleepiness during the night shift in the 6-and-6 plan, and greater within-watch increases in sleepiness for the 6-and-6 plan compared to the 4-and-8 plan. "There was a trend toward shorter sleep episodes in the 6-on, 6-off system and sleep was more often split into two episodes."

Also in 2010, though reported later, a four-section, 3-and-9 plan was receiving a positive reception in the U.S. Navy. A subsequent article by CAPT John Cordle of the U.S. navy and Dr Nita Shattuck of the Naval Postgraduate School about the project that appeared in the *United States Naval Institute Proceedings* won the Surface Navy Association Literary Award for 2013 (78). This award recognized the best professional article in any publication addressing surface Navy or surface warfare issues. In that article CAPT Cordle recalled an attempt to use a German four-section 3-3-3-15 plan on *USS San Jacinto* (CG-56). However, while the 15 hours off worked well for the crew, the 3-3-3 period resulted in "a lot of churn at turnover," and the three hours off tended to be wasted time.

Then, a chief petty officer suggested a 3-and-9 plan, and it was a success. This is also the four-section plan suggested by Comperatore and colleagues in 2005 (72, Table 5-3), and the same plan used by the Roman Army, as noted earlier in this book. To implement the 3-and-9:

> The 0000 to 0300 and 0300 to 0600 watchstanders were allowed to sleep in, with berthing areas dark from 1800 to 0900

Quarters was delayed to 1100

An early breakfast and late supper were instituted

Meetings and training were limited to the period, 0900 to 1500

Operations and navigation briefings were delayed to 1530

An afternoon prayer replaced Taps, and Reveille was not sounded (78)

CAPT Cordle noted a number of benefits of the new plan. Circadian rhythms were stabilized (my modeling, described below, suggested a different result), watchstanders were more alert, stable teams were built across departments, and heat stress in hot spaces was reduced.

An extensive news article about the four section, 3-and-9 plan was published in the *Navy Times*. It makes for good reading and is available at the newspaper's website (113).

The U.S. Naval Postgraduate School investigated the 3-and-9 plan (Plan 20). In his 2012 masters thesis, LT Donald Roberts presented an analysis of a four-section, 3-and-9 plan, shown below, which he studied on the cruiser, *USS San Jacinto* (CG 56) in 2010 (268). Instead of dividing the crew's complement into three sections, one divides them into four sections. This approach is quite viable as long as adequate sleeping facilities are available. Roberts reported the initial use of the 3-3-3-15 plan for a month on the *San Jacinto*, as noted by CAPT Cordle, above, and was present for the initiation of the 3-and-9 plan.

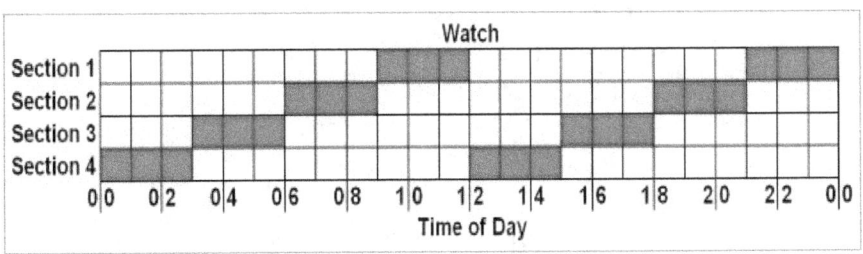

Plan 20. Four-section, fixed 3-and-9 watch plan (Roberts, 2012).

My modeling of section 1 in the 3-and-9 plan predicted an average cognitive performance effectiveness across the watch periods of about 90%, with about 40% of watch time spent BCL. The body clock phase-delayed four hours across 14 days with no stabilization. Three recovery nights of ten hours of sleep were needed. For section 2, the predicted average cognitive performance effectiveness across the watch periods was about 97%, with no watch time spent BCL. The body clock phase-advanced just one hour across seven days, then stabilized. No recovery nights were needed.

For section 3, the predicted average cognitive performance effectiveness across the watch periods was about 95%, with about 20% of watch time spent BCL. The body clock phase-advanced three hours across 14 days, with no stabilization. No recovery nights were needed. For section 4, the predicted average cognitive performance effectiveness across the watch periods was about 94%, with about 20% of watch time spent BCL. The body clock phase-delayed three hours across 14 days, with no stabilization. No recovery nights were needed.

Roberts used a survey to compare the 3-and-9 plan to previous watchstanding experience (Figure 18). There were 104 usable surveys returned, 92 from enlisted and 12 from officers. The breakdown of previous watchstanding experience reported in this survey was instructive, and is reproduced, below. Note that the greatest proportions of experience were with the 5-and-10 (five and dime) fixed plan. This is a clue suggesting widespread use of this three-section plan in the U.S. Navy. CAPT Cordle, above, labeled the five-hour watch as "traditional."

Previous Experience	Enlisted	Officer
5-and-10	85%	83%
5-and-15	60%	33%
6-and-6	87%	50%
4-and-8 fixed	39%	25%
4-and-8 dogged	45%	50%
6-and-12	67%	50%
Other	27%	25%

Figure 18. Previous watchstanding experience reported in 104 surveys, 92 from enlisted and 12 from officers (268). The 5-and-15 plan uses four sections, the 6-and-6 uses two sections, and the other plans use three sections.

The survey comparisons of the 3-and-9 plan to previous experience elicited the following results.

Overall, the crew preferred the 3-and-9 plan to plans in their previous experience.

Enlisted responded that each previous watch type was worse than the 3-and-9 plan except for the four-section 5-and-15 plan, where they were almost evenly split.

About 60% of enlisted reported receiving about the right amount of rest or more on the 3-and-9.

Most of the enlisted said they received more sleep on the 3-and-9 compared to other plans, except for the 5-and-15. About 80% of the 5-and-15 responders reported the same or less sleep on the 3-and-9 plan; about half and half, respectively. [I find the 5-and15 to be a strange schedule: the circadian rhythm must be disturbed greatly by living on a 20-hour day. However, one may certainly acquire more sleep in 15 hours than nine hours.]

Officers preferred the 3-and-9 in a greater proportion than enlisted.

Eleven of twelve officers reported receiving about the right amount of rest on the 3-and-9 plan.

Nine of twelve officers appreciated the 3-and-9 being a fixed plan. [There seems to be a theme in several investigations of officers preferring fixed plans.]

Reviews by Marcus Oldenburg, Xaver Baur and Clara Schlaich of Hamburg's Institute for Occupational and Maritime Medicine supported a couple of observations consistent with the research discussed, above (240). They noted that,

> The duration and the quality of sleep are also important. Sleep requirements and habits may individually vary considerably but everyone requires unbroken periods of rest. According to the Maritime Labour Convention (2006) the maximum hours of work should not exceed 14 h in any 24-hour period and 72 h in any 7-day period. There is a strong association between the number of hours seamen work and sleep deprivation.
>
> Environmental factors such as noise, vibration and adverse weather conditions also disturb sleep quality. (240)

2011

Wilfried Post, Patrick Punte, Peter Rasker, and Anja Langefeld of the TNO had reported the development and testing of an instrument for demonstrating some safety consequences of "innovative manning assignments" (259). In their follow-on study they acquired data across 21 ships, 311 voyages, more than 16,000 watch periods, and almost 60,000 hours of watch time (258). The investigators were examining the possibility of replacing the Chief Engineer with "a beginning maritime officer (MAROF) who has received a dual education in navigational as well as in engineering skills and, as a result, is able to perform tasks in both areas." The data reduction task was quite extensive, and is well documented in the report. The traditional crew was aboard for 177 voyages, while the MAROF was in place for 134 of the voyages. For traditional crews, data were acquired from 84 officers, of whom 20 were Chief Engineers. For MAROF crews, data were acquired from 88 officers, of whom 37 were MAROF. During the watches reported, the ships were underway 75% of the time underway, moored 13% of the time, anchored 7% of the time, and 5% of the time they were arriving or departing (including piloting).

For the Master, the Chief Mate and the MAROF as Mate, there was a statistically significant difference between the two crew types in the number of sleep periods longer than six hours: MAROF ship crews slept over 6 hours more often. There was a significant difference between the two crew types in reported tiredness during their watches: MAROF crews reported about 20% lower tiredness, especially for the Master, compared to the Traditional ship crews, while both crew types reported comparable work loads. However, the average level of tiredness for both types of crews was never more than "a little tired", which does not indicate the existence of fatigue for both types of crews. There were no significant differences between the two crew types for serious shipping incidents or critical situations.

The investigators concluded from their massive amount of data "that sailing with a MAROF combined with shore support is at least as safe as sailing with a Chief Engineer." Further, "on MAROF ships, the [watches] went more satisfactorily and the crews were less tired compared to traditional short sea ships." Short sea trips had been identified as contributing to the fatigue of the navigation officers.

Wessel van Leeuwen, Albert Kircher, Anna Dahlgren, Margareta Lützhöft, Mike Barnett, Göran Kecklund, and Torbjörn Åkerstedt of various Swedish and one U.K. laboratories reported a comparison of "subjective and objective sleepiness in the two most common maritime watch systems: 4h on/8h off versus 6h on/6h off. In addition, the effects of a single free watch disturbance, simulating a condition of low demanding overwork, is compared between the two systems" (169).

Thirty bridge officers participated in a 4-and-8 plan and 19 bridge officers participated in a 6-and-6 plan during a week-long simulated voyage in the North Sea and the English Channel. A simulated disturbance took place between day 2 and 3 or day 5 and 6. "Average sleepiness ratings differed most profoundly between the shift systems during day shifts." During night shifts, sleepiness ratings were higher in the 6-and-6 plan. In both shift plans, sleepiness ratings were significantly higher after the watch disturbance compared to the control condition in the other half of the week. Many work diary components indicated much higher levels of sleepiness in the 6-and-6 plan.

Reaction time did not differ between the shift plans during the night shifts, but was increased moderately during the day shifts in the 6-and-6 plan compared to the 4-and-8 plan. All teams except the 4-and-8 during the 0000-0400 watch showed increased reaction time after the watch disturbance. More PVT lapses occurred for the 6-and-6 plan. The watch disturbance did not result in increased lapsing, except for the 4-and-8 team in the 0400-0800 watch.

The investigators concluded that subjective and objective sleepiness were higher in the 6-and-6 plan than in the 4-and-8 plan, and that the differences were most profound during the daytime. Also, that overtime work (the disturbance) increased sleepiness in both watch plans.

Looking solely at the 4-and-8 watchstanders, these same investigators determined that both subjective and objective sleepiness peaked during the 0400-0800 watch, "coinciding with a time frame in which relatively many maritime accidents occur" (170). Looking solely at the 6-and-6 watchstanders, these same investigators found that "Fatigue and sleepiness levels differed between the watches in the 6-6 shift system with highest levels in the 06-12 watch. This suggests that factors other than circadian rhythms are influencing the results" (85).

Paolo d'Onofrio, Helena Petersen, Johanna Schwarz, Branimir Pantaleev,

Davinder Sharma, and Torbjörn Åkerstedt of Stockholm University and of Southampton Solent University in the U.K. reported on a similar simulation study of "sleepiness among deck officers and engineers working continuous 7 day simulated voyages with normal navigational challenges using the traditional two-watch (6 h on/6 h off) system" (242). The issue of concern was that the use of only two bridge officers with alternating watches on the bridge was becoming more frequent since it was considered to be economically favorable. "This may lead to serious threats to sufficient wakefulness on the bridge."

Fourteen officers stood the noon and midnight watches, and another 14 stood the 1800 and 0600 watches. The investigators used the Karolinska drowsiness test (KDT; EEG alpha & theta power) to detect sleepiness during the first and last two days of the week (148, 260).

Sleepiness increased significantly from first morning to the last morning, but not from the first evening to the last. The group standing the 0600/1800 watches were sleepier in the morning than in the evening. The investigators concluded that seven days spent on the 6-and-6 plan had "a cumulative negative effect on sleepiness levels" that was evident only in the morning.

These latter reports were of preliminary findings from Project Horizon, the final report of which is summarized next.

2012

Project Horizon was an EU-sponsored research project managed by the Warsash Maritime Academy at Southampton Solent University. Participants included the Bureau Veritas of France; Chalmers University; the European Transport Workers Federation (Nautilus) in the U.K.; Stockholm University's Stress Research Institute; the Charles Taylor & Co P&I Club in the U.K.; the European Community Shipowners Associations (ECSA) in Belgium; the European Harbor Masters Committee (EHMC) in The Netherlands; the International Association of Independent Tanker Owners in the U.K.; the U.K. Marine Accident Investigation Branch (MAIB); and the U.K. Maritime and Coastguard Agency (MCA).

Project research focused on how the different watch plans influence the fatigue or sleepiness of deck and engine room watchkeeping officers. Many seven day simulations of realistic voyages were completed in 2011.

The data were analyzed by Stockholm University, reviewed by all of the partners and published in a final report titled 'Project Horizon – A Wake-Up Call" (4). The report was available at the web site of the Warshash Maritime Academy. The shift plan-related conclusions cited in the report echo those stated above:

> Overall, more sleepiness was recorded during the first watch of the day, especially among deck teams
>
> Sleepiness was found to increase with time in watch
>
> The off-watch disturbance instantly increased sleepiness
>
> On the whole, sleepiness levels were higher in the 6-and-6 plan than in the 4-and 8 plan
>
> Sleepiness levels did not significantly differ between deck and engine room
>
> Sleepiness levels consistently peaked between 0400 and 0800
>
> Alertness levels consistently peaked between 1400 and 1800
>
> Wake-time diary outcomes indicated better time off following the first watch of the day: rest and recuperation was rated as more efficient and less negative symptoms such as tension occurred
>
> Outcomes got worse during the course of the week
>
> The disturbed watch had adverse effects in both watch plans
>
> Overall, more negative wake-time diary outcomes were reported in the 6-and-6 plan than in the 4-and-8 plan

Other fatigue data and stress level data were measured and reported, also. In conclusion, the investigators stated, "Considering the results of the present study, special attention needs to be paid to:

> The risks in passages through difficult waters in combination with the 6-on/6-off watch system (because of sleep loss)
>
> Night watches
>
> The last portion of most watches (especially night watches)
>
> Watches after reduced sleep opportunity
>
> Individual susceptibility to fatigue also needs to be considered"

These recommendations are certainly in line with the outcomes of preceding research efforts and I heartily endorse them.

The investigators noted that "Data gained from the research is sufficiently robust to provide input to marine-validated mathematical fatigue prediction models within a fatigue risk management system. Overall, it is clear that much of the data gained from the research supports the 'circadian theory' of diurnal performance peaks and troughs and clear evidence of 'sleepiness' risk periods."

LT Matthew Yokeley at the Naval Postgraduate School "attempted to quantify the effects of sleep deprivation on performance and to determine how that performance is changed through the use of the 3/9-watch rotation compared to a traditional four section 5/15-watch" (329). In his NPS master's project, Yokeley employed wrist activity monitors (WAMs), the Psychomotor Vigilance Task (94), a validated sleepiness scale, and a validated sleep quality scale. Data were acquired while underway on the destroyer *USS Jason Dunham* (DDG 109). Two 13-day periods of data collection occurred, separated by 37 days. The first period was a baseline period, while the second was a test period. During the baseline period, participants worked either a three-section 5-and-10 (five and dime) plan or a four-section 5-and-15 plan. During the test period, participants worked either an alternative, three-section 4-and-8 plan or an alternative, four-section 3-and-9 plan.

Yokeley recovered comparative WAM data for eleven participants. Only one of these eleven acquired more than seven hours of sleep per day during the baseline period (5-and-15). During the test period (3-and-9), two participants acquired more than 7 hours of sleep per day.

In another examination of data from the *Dunham*, a preliminary analysis indicated that sailors working the 3-and-9 slept 86 minutes per day more than others working the 6-and-6 (287).

Michel Paul, Daniel Ebisuzaki, Jason McHarg, all of Canada's Defence R&D Centre—Toronto, Steve Hursh and I evaluated several watch plans used aboard *HMCS St John's* (FFH 340) at the end of Operation Nanook 2011, over the eight days that the frigate transitioned from the high Arctic to Halifax. This evaluation was conducted in light of the watch plan modeling that we had done for two-section submarine operations (249). Based upon that modeling, discussed in the submarine section, below, we noted:

> The watch schedule used by the tactical sailors in [the Canadian] surface fleet is 7-5-5-7. This means that the sailors work a total of 12 hours each day but rather than one main sleep period, they have two shorter sleep periods. We believe that the 7-5-5-7 watch used by our surface fleet is better than the original 6-6-6-6- watch used in our submarines. However, from a fatigue management perspective, the current surface fleet watch system it is not as good as the new submarine watch (8-4-4-8). ... The results of this trial will inform our thinking as to possible improvements in the CF surface fleet watch schedule.

We collected wrist activity data from forty-five sailors. Ten were non-watch-standers, 14 were from the 1 in 2 Port (Front) watch, 14 were from the 1 in 2 Starboard (Back) watch, three were from the 1 in 3 Engineering watch, and four were from the 1 in 4 Engineering watch. Actigraphically-measured sleep and daily work hours were inputted to FAST to generate modeled cognitive effectiveness for each subject. Additionally, the subjects maintained a daily activity, sleep and mood log.

Modeled cognitive effectiveness showed predicted performance equivalent to intoxicated levels of blood alcohol (BAC 0.05% and 0.08%) and worse for all watch periods. The 1 in 2 Port (Front) watch reported less difficulty getting to sleep relative to the non-watch standers. The non-watch-standers reported being in a 'happier' mood than either of the 1 in 2 Port (Front) and Starboard (Back) watch. For all watchstanders and non-watchstanders combined, six Sustained Operations Assessment Profile (SOAP) parameters deteriorated during the trial relative to the pre-trial baseline. These were difficulty concentrating, level of depression, level of irritability, level of fatigue, work frustration and physical discomfort. We concluded that the surface fleet watch plans in use were not optimal.

In our recommendations we indicated that the best watch schedule change for the sailors who stood the 1 in 2 watches would be to move to a straight 8s, 1 in 3 plan where a single 8-hour watch is worked each day. "This would leave 16 hours for meals, training and personal administration and an 8-hour time in bed. Further the three 8-hour work periods should be 1200-2000, 2000-0400 and 0400-1200. This would distribute the work during the WOCL [window of circadian low] across two watch syndicates."

Alternatively, we indicated that they might replace the 1 in 2 watch plan with the 8-on, 8-off, 4-on, 4-off plan (249) and that, while not optimal, this would be a significant improvement over the 7-on, 7-off, 5-on 5-off, 1 in 2 watch plan then in use.

We noted that the engineering watches used seven work periods, including two 2-hour dog watches, ensuring that the engineering personnel worked different hours from one day to the next. "From a circadian/fatigue management perspective, this approach is counter-productive in that sailors do not get use to working the same hours each day and thus cannot adapt to their work hours from a circadian rhythm point of view." As an alternative we suggested using six work periods (0000-0400, 0400-0800, 0800-1200, 1200h-1600, 1600-2000, 2000-0000) so that each engineering sailor would work the same two 4-hour work periods each day (249).

In a master's thesis produced at NPS, LT Stephanie Brown collected data from 15 of the 40-person crew aboard a U.S. Navy combatant surface ship (33). Each subject wore the Actiwatch brand wrist activity monitor (WAM) from Respironics, Inc. They filled out logs, and performed a cognitive attention-switching task (manikin vs. single-digit mental arithmetic) included in the ANAM battery (265), and they performed PVT tests. The WAM data were processed through the FAST software. FAST-predicted performance was compare to actual performance. [I note here that SAFTE and FAST were not designed using PVT data. They were designed using task data such as that acquired from the more complex ANAM switching task. A PVT lapse equivalency was added to FAST after the fact (137). Thus, I hesitate in general to lend much credence to comparisons of FAST predictions to PVT performance.]

Some of Brown's findings were as follow.

> Sleep quality was reduced during underway periods due to an increase in activity during sleep periods. This finding was expected due to increased motion in the sleep environment.
>
> Overall daily sleep quantities were seen during higher sea states. This finding may be due to the poor sleep quality achieved during high sea states, requiring additional overall time in bed to compensate to reduce fatigue symptoms as reported in the surveys.

> Overall performance decreased with ship motion, but stabilized at high sea state.
>
> When actual performance increased, predicted performance either stayed the same or decreased, indicating that the SAFTE model underestimates performance. The model could be overly sensitive to sleep disturbances or poor sleep quality. The finding indicates that the SAFTE model is accounting for the decrease in sleep, but overestimates the effects of poor quality sleep, or disturbances in sleep due to motion.

It's difficult to judge the reliability of this last statement for two main reasons. First, the PVT was not used to develop SAFTE and FAST. Second, the switching task had a noticeable learning curve that extended from pre-cruise, in-port testing throughout the cruise. Holding these two issues in mind, I could not find adequate justification within the Results section of the report for the final statement.

2013

Marcus Oldenburg, B. Hogan and H.-J. Jensen of Hamburg noted the following in a review of 109 studies on the stress and strain experienced by seafarers conducted during the period January 1990 to January 2012 (241, abstract only):

> Only 13 of the identified maritime studies were conducted as field studies, and in 10 of these studies, the focus was on the watch system and/or on fatigue. According to the study results, sleepiness tends to be stronger in the 2-watch system than in the 3-watch system (particularly between 4:00 and 6:00 a.m.). ... Fatigue does not appear to depend on the seafarers' age and is often associated with poor sleep quality; noise and night shifts are also considered to contribute to fatigue.

These reviewers also noted, "Occasional short sleep episodes appear to provide adequate recovery." That observation runs counter to the results of a number of investigations cited above.

Again, Wessel van Leeuwen, Albert Kircher, Anna Dahlgren, Margareta Lützhöft, Mike Barnett, Göran Kecklund, and Torbjörn Åkerstedt of various Swedish and one U.K. laboratories reported a study of the classic, fixed, 4-and-8 plan that included one episode of missing evening or nighttime sleep between two watches (168, 169). Thirty bridge officers participated in a simulation of a seven-day voyage in the North Sea and English Channel. Each participant worked in one of the three watch sections. Half of the participants worked straight through the eight hours between the second watch of day two and the first watch of day three, while the other half did the same thing from days five to six.

The participants provided a sleepiness rating once per hour and performed a five-minute test on the Psychomotor Vigilance Task (PVT) at the beginning and end of each watch. The investigators recorded the sleep EEG for polysomnography during six watches in the first and the second half of the week.

Sleepiness was higher during a section's first watch of the day watch than during their second watch. In addition, sleepiness increased with hours on watch, peaking at the end of the watch. The non-sleep period increased sleepiness strongly. Similarly, PVT reaction times were slower during the first watch of the day than during the second, and slower at the end of the watch than at the beginning. And similarly, the non-sleep period increased reaction time strongly.

The average daily total sleep times were 7.4 hours for the 12 to 4 watch section, 6.6 hours for the 4 to 8 watch section, and 6.9 hours for the 8 to 12 watch section. The participants generated about 5.5 hours of sleep in the sleep periods that occurred from midnight to 0800 and 0400 to noon. About four hours of sleep occurred in the 2000 to 0400 sleep period. The daytime sleep periods allowed about 1.5 to 2.5 hours of sleep each. Polysomnography caught one-third of the participants falling asleep on watch, especially during the midnight to 0400 watch. Falling asleep on watch increased after the non-sleep period.

The investigators concluded, "subjective and objective sleepiness peaked during the night and early morning watches, coinciding with a time frame in which relatively many maritime accidents occur. In addition, we showed that overtime work strongly increases sleepiness. Finally, a striking amount of participants fell asleep while on duty."

LT James Davey in his NPS master's project assessed "the relationship between fatigue and crew member performance on the Psychomotor Vigilance Test and the Switching Test" (88). I liked Davey's opening quotation:

> Why when I PCS [transfer between permanent

> duty stations] does the government want me to only drive 350 miles and get 8 hours of sleep but when I drive a BILLION DOLLAR Warship I only need 4?
>
> — From the unofficial Junior Officer Protection Association (JOPA) Facebook Page

Also, Davey's own observations,

> From the founding of the U.S. Navy and the ships of John Paul Jones and Commodore John Barry, the Navy has been a place in which the term "sleep when you're dead" has been ingrained in the culture of surface warfare officers and the sailors under their command. Guzzling coffee and "manning up" became the tradition, and the corporate knowledge of how to manage watch and work schedules was passed from one generation of sailor to the next with no thought or worry as to the long- or short-term effects on a sailor's health or cognitive and physical readiness. Notwithstanding the incredible technological advances that the Navy has achieved since its founding on October 13, 1775, the demands on a sailor's time and the need to be awake and alert have not wavered. The need to balance work, watch, training, drills, meetings, and sleep have always come at the expense of a sailor's sleep schedule and the Navy's combat readiness.

Davey provided a compact summary of total sleep per 24 hours observations made by master's students who preceded him at NPS. I have copied his table 1, below. I have discussed these studies, above. The means and standard deviations in the table imply that in the Nguyen 2002 study (234) total sleep was only 4.78 hours and less in 5% of cases, and in the Mason 2009 study (185), only 3.66 hours and less about 5% of the time. So much for the safe operation of the billion-dollar warship!

Author and Year	Ship Class	Ship Employment	Average Hours of Sleep + (STDEV)
Nguyen (2002)	CVN	Op Enduring Freedom	6.28 (1.50)
Haynes (2007)	DDG	Training	7.27 (1.03)
Mason (2009)	CG x 2	Training	5.58 (1.92)
Green (2009)	FFG	Training	6.71 (NR)
Brown (2012)	LCS	Training	7.36 (2.63)

Figure 19. Total sleep per 24 hours observations made by master's students at NPS (88).

Davey's own data were from most of the 40 crewmembers operating the test littoral combat ship *USS Independence* (LCS-2). The data were from a ten-day test and evaluation trial of this new ship type in May 2013. Davey failed to specify the watchstanding plan used during the test. From another source, I assume that two submarine fleet procedures were being applied to the development of the LCS concept (144). One was the rotation of blue and gold crews on and off of the vessel. The other was the use of the three-section 6-and-12 watchstanding plan.

As a fatigue scientist, I cannot think of a worse idea for a watchstanding plan for a surface vessel. It has been used extensively in U.S. submarines as discussed, below. However, this practice was based upon flawed, older assumptions about the human circadian rhythm. I have also discussed this problem, below. Why in this modern era would you try to impose an 18-hour wake-sleep cycle on crewmembers exposed to a natural 24-hour cycle of daylight and darkness? This idea flies in the face of everything that we now know about human sleep and circadian rhythms. Perhaps the idea was that only a limited amount of cumulative fatigue would build up for a single team across a ten-day mission. Davey's data indicated otherwise.

Davey conducted 13 correlational analyses, including as predictor variables total sleep time, sleep efficiency, and FAST predictions of cognitive effectiveness, with cognitive performance measures as dependent variables. Davey stated that none of these relationships was statistically significant. This statement was a bit inaccurate.

Taking a look at Davey's regression results for the slope of the predictor line (Davey's table 5), we see that daily sleep efficiency predicted response speed on the PVT at the 91% level of confidence, switching test reaction time (RT) at the 99.9% level of confidence, and switching test throughput at the 95% level of confidence. The problem with these significant slopes was that they did not explain a useful amount of the variance in the data, with shared variance values ranging from 2.4% to 12.4%. So, while they were statistically significant results, they were not useful and Davey discounted them with good reason.

The sequence of test days proved useful as a predictor of cognitive performance at first blush. The sequence of days explained about 31% of the variance in switching test RT and about 43% of the variance in switching test throughput, and both of these regression slopes were statistically significant at the 99.9% level of confidence. However, these were just learning curves (decreasing RT and increasing throughput across days). Davey showed that the learning curve for the switching task flattened at day 7.

Davey also conducted six tests for differences across days for sleep time, sleep efficiency and cognitive performance measures. Sleep duration was significantly shorter on day 5 than on day 1, with day 5 being the low point across the ten-day test.

Thirty-one of the subjects wore wrist activity monitors (WAMs) during the test. Davey used the sleep estimates from the WAMs to determine whether the subjects achieved the 480 minutes of daily sleep specified in the Navy Standard Work Week (NSWW). Few did. The average daily sleep deficit was 1.8 hours, compared to the NSWW specification.

Limiting himself to the RT issue, Davey compared the known speeds of anti-ship missiles to the fatigue-related RT impairments of sailors manning counter-ordnance. A fatigued four-person crew could experience a total of a three-second RT delay, allowing a missile to come one or two miles closer to the ship than normal before counter-ordnance would be fired. A very interesting and practical analysis for the military! Davey went on to place his findings within the context of error and cognition models.

LT Roger Young at NPS reported a comparison of the sleep and performance of sailors standing the four-section 3-and-9 plan or the two-section 6-and-6 plan on a deployed warship (331). Data were collected for twelve days aboard the *USS Jason Dunham* using WAMs, sleep and activity diaries, and response speed on the PVT. [Response speed is the reciprocal of response time, and for the PVT is more normally distributed for statistical analysis.] Thirty-three participants, of whom 20 were enlisted, provided full or partial data.

The average daily sleep amount for participants on the 3-and-9 plan was 6.5 hours, while the 6-and-6 participants slept 5.5 hours per day. The 3 to 6 section in the 3-and-9 plan generated the most daily sleep at 7.1 hours, the 6 to 9 section generated the least at 6.1 hours, and the two sections of the 6-and-6 each generated about 6.5 hours.

The average response speed for participants standing the 6-and-6 plan was 3.66, while the 3-and-9 participants had a higher average speed of 3.94. Response speed increased during the first week for the 6-and-6 watchstanders, and decreased for the 3-and-9 watchstanders, becoming about the same speed for both groups in the second week.

Young concluded, "[S]ailors standing the 3-and-9 watch rotation, on average, performed better on the PVT than sailors who were standing the 6-and-6 rotation and receiving less sleep."

2014

Nita Shattuck, Panagiotis Matsangas and Lauren Waggoner of the NPS reported the results of a comparison of the four-section 3-and-9 plan (Plan 20) to a dogged, four-section, 6-and-18 plan in which the 0000 to 0600 watch was dogged (Plan 21) (281, 286). The latter data were from 34 crewmembers on the *USS Jason Dunham* (DDG 109) an Arleigh Burke Class destroyer, deployed near the Persian Gulf. The 3-and-9 data came from an unpublished data collection effort of December 2012 on the crew of a similar ship.

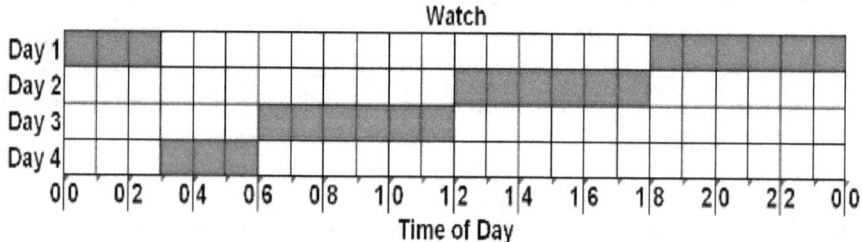

Plan 21. Four-section, counterclockwise-dogged, 6-and-18 watch plan in which the 0000 to 0600 watch is dogged (286).

My modeling of Plan 21 predicted 90% cognitive performance effectiveness across watch periods, with 35% of watch time spent BCL. The circadian rhythm had a one-hour peak-to-peak oscillation with a four-day period, and recovery was immediate.

The investigators' results indicated that the fixed 3-and-9 plan (Plan 20) was superior to the dogged 6-and-18 plan (Plan 21). My modeling was in agreement with this assessment. The 3-and-9 allowed 0.63 more hours of rest daily, 0.89 more hours of sleep daily, and decreased variability in psychomotor vigilance performance, both reaction time and lapses. The variability of daily rest and sleep was also reduced. They concluded that the 3-and-9 plan "yielded better sleep hygiene, more stable performance and was well-accepted by crewmembers." Also, "This study suggests that watchstanding schedules based on sound human performance and ergonomics principles may lead to better performance in the operational environment that are better accepted by crewmembers."

2015

Nita Lewis Shattuck, Panagiotis Matsangas, and Stephanie Brown at the Naval Postgraduate School compared the 3-and-9 watchstanding plan to the 5-and-10 watchstanding plan (283). Data were acquired from 117 belowdecks crewmembers from the Reactor Department on the USS Nimitz (CVN 68). Sleep data were acquired with Motionlogger Watch (Ambulatory Monitoring, Inc., Ardsley NY) and scored with the Cole-Kripke algorithm (53). Mood and sleepiness ratings were acquired. Performance data were acquired using the PVT, and performance predictions were made with the FAST software. The investigators concluded that:

> The 3-and-9 plan was better than the 5-and-10 in terms of sleep quality, subjective levels of fatigue, mood, psychomotor vigilance performance, and acceptance by the Sailors. Mainly, the sleep hygiene and acceptance of the two schedules differed considerably.
>
> Compared to their counterparts on the 3-and-9 plan, crewmembers on the 5-and-10 plan had 15% longer reaction times and 59% more lapses and false starts on the PVT.
>
> Over a 3-day rotation cycle, a crewmember on the 5-and-10 plan slept at three distinctly different times on each subsequent day, experienced two periods of sustained wakefulness, and had one night with only a 4-hour opportunity for sleep. In contrast, sailors on the 3-and-9 plan had a fixed sleep schedule every day and they experience, at most, 16- to 17-hour periods of sustained wakefulness.
>
> Two 3-and-9 night watches sections were accepted poorly by the sailors. The investigators suggested that these concerns could be ameliorated by adjusting sunlight exposure, providing sleep hygiene training, and considering an alternative distribution of

duties between watch sections to allow protected sleep times for those on night watch duty.

In another study, Shattuck and Matsangas also used "actigraphy and the Fatigue Avoidance Scheduling Tool (FAST) to quantify the sleep patterns of a 39-year old Commanding Officer (CO) of an Arleigh Burke class destroyer while the ship was forward-deployed" (282). This study echoed concerns voiced by psychiatrist Jonathan Shay at the U.S. Army War College about the amount of sleep acquired by commanders, whose decision processes affect the lives of those whom they command (288). Shattuck and colleagues summarized their results as follow. "On average, the CO received 5.2 hours of sleep daily and averaged 6 hours in bed each day. The participant received more than eight hours of sleep for only 2% (n=3) of the study days; for 17% (n=27) of the days, he received less than 4 hours of daily sleep. For 15% of waking time, the CO had a predicted effectiveness of less than 70% on the FAST scale, equating to a blood alcohol equivalent of 0.08% -- or legally drunk. The CO's predicted effectiveness was below 65% approximately 10% of waking time."

The investigators pointed out,

> [S]leep deprivation impairs decision making involving the unexpected, innovation, revising plans, competing distraction, and effective communication [(123)]. Restricting sleep to approximately 5 hours for seven consecutive nights has been shown to have a significant negative effect on alertness in terms of sleepiness, fatigue, mood disturbance, stress and psychomotor vigilance performance [(93)].

They also noted,

> [T]he additional drop in daily sleep observed after a critical event had occurred raises the question of limitations of spare capacity. If the CO is already chronically sleep-deprived, when a critical event occurs, his/her ability to react may be undermined by the chronic sleep restriction prior to the event. Moreover, the critical event itself often leads to further reductions in sleep, becoming a "vicious circle" of degraded performance, especially if the critical events are clustered in time. In such a case, operational commitments may not allow for adequate restoration of sleep reservoirs since operational environments have even less opportunities to get recovery sleep.

Shattuck, Matsangas and Edward H. Powley at NPS reported an investigation of "sleep patterns, mood, psychomotor vigilance performance, and command resilience of watchstanders on the 'five and dime [5-and10] watchbill" (285). The investigators acquired data across about two weeks in port and underway from 77 sailors who worked in the Reactor Department of the *USS Nimitz* (CVN 68). The instruments used by the investigators included diaries for sleep quantity and quality, daytime sleepiness, sleep conditions; wrist activity monitors; psychomotor vigilance performance (PVT); Fatigue Avoidance Scheduling Tool (FAST) predicted effectiveness; and assessment of organization commitment, psychological safety, and command resilience in the RX Department. The latter was accomplished with the Command Resilience Questionnaire (CRQ), the Profile of Mood States (POMS), the Organizational Commitment Questionnaire (OCQ), and the Psychological Safety Questionnaire (PSQ).

> Although crewmembers on the 5/10 received approximately seven hours of sleep per day, they reported experiencing excessive fatigue and dissatisfaction with the schedule. This contradiction is best explained by examining sleep and rest periods over a 72-hour period, during which a crewmember sleeps at three distinctly different time periods each day. On the first day of the cycle, the Sailor typically receives an early-terminated 4-hour sleep episode followed by two periods of sustained wakefulness, 22 and 20 hours. During these periods, daytime napping only partially ameliorates the fatigue and sleep debt accrued during these periods of sustained wakefulness. Given this pattern, it is not surprising that at the end of the underway phase, the crewmembers' moods had worsened significantly compared to moods at the beginning of the underway period. Psychomotor vigilance performance in the 5/10 is comparable to the performance of Sailors on the 6hrs-on/6hrs-off (6/6) schedule [see previous NPS reports, above]. It is significantly degraded compared to Sailors on the modified 6hrs-on/18hrs-off (6/18) and the 3hrs-on/9hrs-off (3/9) schedules. Specifically, the 5/10 had 21.4% slower PVT reaction times, and 71.5% more lapses plus false starts than the 3/9. Our findings suggest that the 5/10 watch, combined with other work duties, leads to poor sleep hygiene. Crewmembers on the 5/10 suffer from sustained wakefulness because of extended workdays and circadian-

misaligned sleep times. In general, the self-reported survey results suggest low degrees of resilience, psychological commitment to the organization, and psychological safety. In terms of organizational commitment, participants report that they do not talk positively about their department and do not view their department as inspiring performance. Conversely, Sailors report a high degree of willingness to put in effort beyond expectations, even though overall results indicate low psychological attachment to the unit as a place for working and completing work tasks. Results also show low levels of psychological safety. (Abstract)

Surface Watchstanding Research Summary

There is no doubt that the requirements to work at night instead of sleeping, and to sleep during daylight hours instead of during the night, impair cognitive performance and decision-making in most people, including seafarers. There is also no doubt that the effects of these requirements elevate the risks of accidents and untoward incidents. The degree of elevation varies both within and across different industries and different modes of transportation. This problem of establishing the degree of elevation usually stems from inadequate post-accident/incident reporting and inadequate post-accident/incident investigative tools. However, from my reading over the decades it seems that relative risks are doubled easily by both acute and cumulative fatigue and may, in the case of nocturnal highway accidents, be increased by at least a factor of ten compared to some periods during the day. Post-accident/incident investigations and reporting may be improved by the use of fatigue-related guidelines (197, 262).

Unfortunately, maritime activities require 24/7 management, as do shore-based emergency services and infrastructure operations. Unlike the latter, in which four- and five-crew shift plans may be used, maritime operations have been limited to two- and three-section plans. The fewer the crews or sections, the greater the work demand placed upon a crewmember or watchstander in terms of the numbers of hours worked per day and per week and, often, the numbers of hours allocated for nocturnal sleep (196).

None of the watchstanding plans discussed, above, is a _good_ plan for fatigue management. There is no _good_ plan for any kind of night work that prevents sleeping during the midnight to dawn period. Instead, each plan is an attempt to support the demand for 24/7 operations and, in some cases, to minimize the effects of concomitant human fatigue upon system efficiencies and safety.

Chapter 8. Research Belowdecks

This is a transitional chapter between my chapters dealing with surface operations and those dealing with submarine operations. The issue considered here is the limited exposure to daylight of certain crewmembers of surface vessels. In contrast to surface operations, submarine crews tend to work and sleep in low-level lighting conditions, separated from strong photic *Zeitgebers* (135). Hunt and Kelly measured light levels aboard the nuclear submarine *USS Asheville* (SSN 758). While lighting levels varied widely, they were generally below the levels thought then to serve as a *Zeitgeber*. They pointed their meter at the light in an officers' pantry and measured 1188 lux, the highest onboard reading. The crew's mess varied from 97 to 324 lux, the engine room from 76 to 540 lux, and the sonar and navigation areas from 0 to 206 lux.

When this study was conducted, it was thought that these light levels were too low to entrain the phase of the circadian rhythm. Thus, the investigators concluded that free-running of the circadian rhythm might occur in submariners. However, subsequent research has indicated that the circadian rhythms of people living in very low-light environments may be entrained by pulses of very low levels of light such as observed onboard *Asheville* (99).

Now, consider crewmembers who work belowdecks on surface vessels. LT John Nguyen and Dr. Nita Miller Shattuck at NPS examined the effects of reversing the work-sleep schedules of the crew aboard the aircraft carrier, *USS John C. Stennis* (CVN 74), involved in night combat operations in Operation Enduring Freedom (215, 234). [Note that Kleitman presented a complete history of older research concerning wake-sleep schedule inversion (155, pp. 173-174).]

Crewmembers on *Stennis* arose at 1800 for breakfast and worked through the night and early hours launching and recovering aircraft until 1000, when their duty day concluded. Nominally, they worked a 12-and-12 watch plan, but the nighttime "workday" was a bit longer. Twenty-eight enlisted crewmembers provided 72 hours of actigraphy. Data collection occurred at the end of *Stennis'* 30-day assignment to night operations, by which time some acclimatization to night work might be expected.

Twelve of the 28 participants worked topside, and the remainder belowdecks. Belowdecks participants averaged 7.4 hours of sleep per day while those working topside averaged only 4.7 hours. About 75% of the participants working belowdecks received more sleep on average than anyone working topside. The participants were asked, "Have you ever completely adjusted?" to the inverted work-sleep plan. Thirteen of the 28 answered 'No'. Nguyen concluded that a significant number of sailors had difficulty adjusting to working nights and sleeping days, and that those working topside had greater difficulty adjusting to the reversed schedule than those working belowdecks.

According to Ensign Tiffoney Sawyer, during this same project (278),

> The results showed that younger participants were angrier than older participants on night shiftwork. The results also indicated that there was a significant interaction between repeated measures of mood states and gender. In addition, female participants reported significantly higher mood scale scores than the male participants, and topside participants were getting significantly less sleep than belowdecks participants.

It was reported to me in 2014 that some engineering departments use a fixed 8-and-16 plan, and that this may be due more to tradition and organizational resistance to change than to watch plan analysis. For belowdecks workers, this is not a bad plan, given the use of bright light therapy and bright light protection. Regarding light therapy, the following research paper may be of interest, though it was not conducted with mariners.

The Use of Bright Light Therapy

The researcher literature on this subject is quite extensive. I highlight here a single study, leaving the reader to investigate further. Western Michigan University conducted an investigation to determine whether blue light therapy might "improve behavioral alertness in flight crew-members" (32). The investigators provided a readable and useful review of the light therapy literature. In their experiment, they collected both subjective and objective data for a month, using well-validated research instruments, from fourteen flight crew members working as pilots or flight attendants. The first two weeks served as a baseline. This was followed by daily, 30-minute blue light (465 nm) interventions, conducted in flight during normally-scheduled long haul flights. They found significant improvements in subjective sleepiness scores, subjective fatigue scores, and performance on the psychomotor vigilance task (PVT) as a result of light treatment.

This kind of study needs to be repeated using maritime watchstanders on underway vessels. Enough research on the benefits of light therapy for the mitigation of fatigue has occurred that Brown and colleagues were able to make the following comments:

> Recently, we have seen a flux of light therapy innovations aimed at passengers to improve mood, decrease effects of jet lag and minimize fatigue. Next generation aircraft such as, the B787 and A380, have mood lighting installed to help passengers adjust to new time zones. Paris Charles de Gaulle airport installed three light therapy 'spaces' where passengers can enjoy light therapy to fight their jet lag. The benefits of light therapy has extended well beyond aviation, and is used with depression, dermatology, psychiatry, neurology and gerontology and work related issues such as, shiftwork and sports (Dutch Olympic Swimming Team, TVM Ice-skating team, Dutch Olympic Committee). We have also seen light cafes opening in Seattle and Sweden. An energy company in Umeå, northern Sweden, has installed phototherapy lights in the city's bus stops to combat the short days, lack of sunlight, and residents' depression.

> In addition to individual crewmember portable light units, crewmembers can benefit from innovative 'light stations' in crew check-in areas. Such light stations have been placed in the operations area at airlines operating in arctic regions and the student lounge area at Western Michigan University, College of Aviation. Students can bask under light therapy while studying or relaxing in between classes. The desktop 10,000 lux, 17,000 Kelvin UV-Free lights mimics a blue sky. In addition to improved alertness, relief from seasonal effective disorder is also a benefit- particularly in the dark winter months in areas such as Seattle, Norway, Sweden, Canada and Michigan.

Certainly, seafarers should be able to benefit from arrangements such as those described here. However, guidance as to when to use phototherapy <u>must</u> accompany the equipment.

The Use of Protective Eyeglasses

The literature on this subject is relatively small. I highlight several studies here. Stephanie Crowley, Clara Lee, Christine Tseng, Louis Fogg, and Charmane Eastman of Rush University Medical Center in Chicago worked with 67 younger subjects around the age of 22 years who worked five consecutive simulated night shifts, 2300 to 0700, then slept at home 0830 to 1530 in darkened bedrooms (83). They wore sunglasses with normal or dark lenses when outside during the day, and also took placebo or melatonin before daytime sleep. During the night shifts, they were exposed to intermittent bright light or remained in dim light. The sample was split into an 'earlier' group, whose body temperature minimum occurred before 0700, and the concomitant 'later' group. "The later participants were completely re-entrained regardless of intervention group, whereas the degree of re-entrainment for the earlier participants depended on the interventions." "With only room light during the night shift, darker sunglasses helped earlier participants phase-delay more than normal sunglasses, but melatonin did not increase the phase delay." The investigators recommended the combination of intermittent bright light during the night shift, sunglasses (as dark as possible) during the commute home, and a regular, early daytime dark/sleep period if the goal is complete circadian adaptation to night-shift work.

Alexandre Sasseville, Nathalie Paquet, Jean Sévigny and Marc Hébert of Laval University in Quebec examined the effects of blue blocker glasses on nighttime melatonin secretion when subjects were exposed to bright light at night (277). They exposed 14 subjects to a 60-minute bright light pulse between 0100 and 0200 while the subjects wore orange (blue blocker) lens or grey lens glasses. A small, 6% increase in nighttime melatonin production was observed with the orange lens reduction of 46% reduction of nighttime melatonin secretion was observed with the grey lens. Thus, the investigators concluded, these blue blocker glasses were "an elegant means to prevent the light-induced melatonin suppression."

Kimberly Burkhart of the University of Toledo, Ohio, and James R. Phelps of Samaritan Mental Health in Corvallis, Oregon, had 20 subjects wear "either blue-blocking (amber) or yellow-tinted (blocking ultraviolet only) safety glasses" for three hours prior to sleep (37). The subjects "completed sleep diaries during a one-week baseline assessment and two weeks' use of glasses." Sleep quality was reported to be poorer in the amber lens than the control group at the beginning of the study outset; one of the hazards of assigning subjects to groups randomly. By the end of the study, the amber lens group experienced a statistically significant improvement in sleep quality relative to the control group and their mood was improved significantly compared to the beginning of the study and relative to controls. These data support the practice of protecting yourself in the evening against the suppression of melatonin suppression by bright light.

Stéphanie van der Lely, Silvia Frey, Corrado Garbazza, Anna Wirz-Justice, Oskar Jenni, Roland Steiner, Stefan Wolf, Christian Cajochen, Vivien Bromundt, and Christina Schmidt of the University of Basel, University Children's Hospital of Zürich and Ilmenau University of Technology in Germany studied thirteen teenagers, 15 to 17 years old (171). They "investigated whether the use of blue light-blocking glasses (BB) during the evening, while sitting in front of a light-emitting diode (LED) computer screen, favors sleep initiating mechanisms at the subjective, cognitive, and physiological level." They found that "Compared with clear lenses, BB significantly attenuated LED-induced melatonin suppression in the evening and decreased vigilant attention and subjective alertness before bedtime. Visually scored sleep stages and behavioral measures collected the morning after were not modified." They concluded that "BB glasses may be useful in adolescents as a countermeasure for alerting effects induced by light exposure through LED screens..."

Chapter 9. Submarine Watchstanding Studies, 1947-1999

The preceding, transitional chapter, Research Belowdecks, introduced the idea that the body clock might be fooled more easily when the exposure of crewmembers to the natural daylight-darkness cycle is limited or precluded, and the use of bright light or blue light therapy is implemented. In the submarine environment, exposure to the natural daylight-darkness cycle may be limited or precluded as a daily practice to help control the body clock. This chapter (1947 to 1999) and the next chapter (2000 to 2010) deal with research concerning the latter idea.

Additionally, these two chapters chapter deal with watchstanding plans that have been observed in submarine operations. Note that across several western nations, apparently only the United States uses three watch sections on submarines at present (248):

> Unfortunately, only large submarines with large crews ... [may] opt for a 1 in 3 straight eights or straight fours watch system.

> Nations that operate small diesel-powered attack boats which typically have small crews cannot employ a 1 in 3 straight eights watch system. Small diesel-powered boats are limited to a 1 in 2 watch system where they are on watch for 12 hours a day.

1947

To set the context for much of the research discussed subsequently in this chapter, I have pulled some excerpts from pages 99-114 of an historical document concerning U.S. submarine operations in World War II. Commander (Dr.) Ivan F. Duff (MC), USNR, provided information about "human factors" that were noted in 1,471 of the approximately 1,500 war patrols made by U.S. submarines in the period 7 December 1941 to 14 August 1945 (97). Most of these patrols were in the Pacific, and most were in the larger "fleet" submarines as opposed to S-boats. This article included excellent descriptions of habitability issues, which I do not address in this book. The habitability issues included especially air quality, heat, cold, food, and water.

> This was an existence characterized by very crowded living and sleeping conditions, limited water supply, and frequent high

temperatures and humidity resulting from engine room heat and shutting down ventilation during certain periods of enemy contact.

Many missions were marked by days of fruitless patrolling and almost unbearable monotony and boredom, sometimes broken by contact with the enemy when excitement and tension were at a very high pitch. Some patrols, though of short duration, were very active, with men remaining at battle stations for hours on end.

Sleeping accommodations were so limited that with an average-sized crew of 75 it was always necessary for some of the men to share bunks by sleeping in shifts.

Once the submarine was underway, no one was allowed topside except the authorized watch. An active patrol often necessitated dawn-to-dusk submergence, with the result that sometimes men did not see the sun for days at a time.

Exclusive of men lost on 52 overdue submarines [a euphemism for submarines that were sunk, I presume], only 62 deaths occurred on operations. [Only one was a suicide.]

The success of a submarine's mission was sometimes compromised by defects in personnel health ... Serious or widespread illness was reason for either termination or interruption of approximately 4 percent of all patrols. The cause of termination of 29 patrols on the basis of illness has been analyzed as follows: [The number one cause, occurring in nine patrols, was "Excessive personnel fatigue."]

Personal fatigue of this magnitude occurred only in the first two years of the War, terminating patrols of eight fleet-type submarines. In five other instances personnel endurance was exhausted and would have terminated the cruise had not operation orders done so.

Bunking facilities aboard many submarines were designed for peacetime complements which were increased as the war progressed. ... Overcrowding on the average patrol probably produced no lasting effect on personnel efficiency. On submarines used in Air-Sea Rescue Operations and as troop transports, serious overcrowding was experienced. In preparing for the future, the use of submarines for such purposes merits review...

In 1941 and 1942 little was known concerning the length of war patrols men might be expected to tolerate. ... At the time of leaving the area of concentrated activities an attempt was made by commanding officers to estimate the remaining days of personnel endurance. This figure was arrived at subjectively. Though it was reported as "0" days on 25 patrols, this did not necessarily mean that the crew was in a state of collapse, but rather that the men were no longer on their toes and that their fighting efficiency had dropped. Early evidence of fatigue was commonly observed about the 40th to 50th day on station.

A patrol carried out in good weather with plenty of targets, with good fire control, and without depth charges could last for longer than one on which any one of these features was missing. The monotony of a submerged patrol without contacts was very fatiguing unless some change of pace or diversion was introduced. Short aggressive patrols were said to take as much or more out of the men than did the longer patrols. If lulls in activity occurred, material reduction in efficiency could be forestalled. If not, fatigue began to be apparent. Although aggressiveness and desire to close with the enemy might not have slackened, the fighting edge of the ship was definitely impaired, in that the reserve strength of the crew to meet possible emergencies was lacking. Failure to "shift to the second string" under such circumstances was sometimes cause for regret. On these occasions fatigue sometimes became dangerous, endurance approaching near physical and nervous exhaustion. The last week of the patrol was generally the hardest. Recuperation during the quiet return voyage was often noticeable and sometimes acted to create a false impression of the crew's endurance upon arrival at port. When fatigue was excessive, recuperation might not take place, especially if considerable action had been experienced in the last week on station.

Extensive surface operations, while inherently decreasing the amount of rest, increased ·the general well-being of the crew. On prolonged submerged operations a routine was necessary which would keep efficiency at the highest level. Continual rough seas with "colds," seasickness, need for securing egress of air from the outside and inability to sleep produced a very depressing effect. Under such circumstances, withdrawing from station for a short rest was authorized.

In a similar review of patrol reports, Commander (Dr.) Duff and Captain (Dr.) Charles. W. Shilling (MC), USNR, assessed psychiatric casualties in wartime submarine operations (98, pp. 117-128). CAPT Shilling was a founder of the Naval Submarine Research Laboratory in Groton, Connecticut (87).

There can be no doubt that the trauma sometimes experienced by personnel in the Submarine Service was as great, if not greater, than that experienced by any other group in the war. With every depth charge attack the officers and men could not help but wonder when the next one would make a direct hit. They all knew that submarines were being lost to enemy counter-attack.

"For the first two hours we were in a mighty tough spot. Extreme dis.. comfort was suffered from the accumulated heat and humidity. All hands stripped down to shorts and the men took off their shoes and socks... the predicament of the ship was a fact fully recognized by the older and more experienced men. As the youngsters folded up, the others took over. The most startling effect was the apathy engendered by the combination of heat, pressure, physical effort and mental stress. Some without permission, others after requesting relief, would seek the closest clear space on the deck, lie down, and fall asleep. Most stations ended up with two men taking turns, relieving one another when necessary, the off-watch resting on the deck beside his station."

The general manifestations evidenced by the men under stress (psychic trauma, physical strain of repairing materiel casualties in excessive heat, humidity and pressure) are described as excessive physical weariness, lethargy, and sometimes heat exhaustion.

The generous use of rest camps, we believe, proved to be the greatest single factor in the excellent record achieved by the submarines. It was the general policy that no man made more than two consecutive patrols without a period in a rest camp. Again, an equally important factor was the rotation to the U.S. to pick up a new boat and of course the incidental leave was an effective form of rehabilitation.

I was intrigued when I read about the use of "rest camps." My father, clinical psychologist Dr. Cecil R. Miller, was the NCOIC of a similar rest camp for pilots in May through September 1944. This camp was the 430th AAFRTU (replacement training unit), a convalescent center for battle fatigue (now called PTSD). It was co-located in Ephrata, Washington, with the 430th Combat Crew Training Station-Standby at Ephrata Army Air Base. I have included one of my father's photos of a camp activity, below.

Figure 20. Comical tree planting "ceremony" at the 430th AAFRTU, Ephrata Army Air Base, 1944. From left, T/4 Cecil Miller, three nurses, another staff member, and two patients. As NCOIC, one of T/4 Miller's duties was to arrange entertaining and amusing activities for the patients. (Photo from the archive of Dr. Cecil R. Miller)

The extracts about WWII submarine operations, above, were taken from the 1947 *History of Submarine Medicine in World War II* (289). This report listed quite a few areas of research interest that were being or needed to be pursued in light of the wartime experiences. I found no mention there of concerns about sleep nor about circadian rhythms.

I mentioned earlier that I had no research reports from maritime powers such as China and Russia. However, I did find a couple of interesting mentions of sleep and watchstanding in WWII German U-boats.

> When German submarines fitted with the snorkel (a retractable breathing tube for procuring air from the surface for engines of a submarine remaining submerged) first became available to us at the close of the war, very little information was available concerning the effect of personnel and living conditions within the boats. Some reports had been made of serious difficulties resulting from the use of this device. Since the snorkel is fitted with a head valve which closes automatically to prevent intake of water when the top of the snorkel tube is ducked, and the sudden closure of this main air intake to the boat causes a sudden and violent change of pressure within the submarine, reports of ear pain and even ruptured ear drums were prevalent, as well as some reports of nausea, disturbance of sleep and other difficulties (289) (p. 203)

> In combat, every man in the crew was on duty, either doing his designated task or standing by to deal with any emergency that might arise. At all other times the crew was divided into watches. The seaman's division worked a three-way system (8 hours on duty, 8 hours of sleep, and 8 hours of miscellaneous duties about the boat) while the engineers worked a two-way system (6 hours in the engine room, 6 hours asleep) and the communications worked a split system with their day being divided into three four-hour watches between 0800 and 2000, and two six-hour watches between 2000 and 0800. When on the surface there was a bridge watch, consisting of an officer and four lookouts. The watch officer was found from the IWO and 2WO, who did two four-hour shifts each day, and the *Obersteuermann* and *Oberbootsmann*, each of whom did one four-hour shift per day. The lookouts were found from the seamen who each did one four-hour shift per day and these men were crucial, since the survival of the boat and all in it depended upon their rapid reactions. (195)

I found it interesting that three different watch plans were used aboard one submerged vessel: 8-and-16 for seamen, 6-and-6 for engineering, 4-and-8 days and 6-and-6 nights for communications. Also, surface watch periods reverted to the classic four-hour length.

1948

As noted at the beginning of my chapters on research associated with surface vessels, Dr. Nathaniel Kleitman of the University of Chicago collected data aboard the *USS Dogfish* (SS 350) in May 1948. This research was performed for the U.S. Committee on Undersea Warfare of the National Research Council (156). As Kleitman pointed out,

> For the purpose of insuring that a portion of the crew is wide-awake and ready for action at any hour of the day and night, the routine that prevails on submarines has several advantages over that of surface ships. There is no reveille and usually no daily quarters requiring the entire crew to turn out. The men are permitted to sleep during daytime hours, and since practically all work is done under artificial light, the difference between day and night is minimized, or absent. (p. 330)

Kleitman's assumption about light effects must now be questioned, based upon the observations of low light serving as a *Zeitgeber*, as mentioned in the preceding chapter (99). In Kleitman's study, the crew operated in three watch sections and one non-watch section, and the sections were each of about the same size, about 17 to 20 men, including two officers each. They used the classic maritime fixed watch plan (Plan 1). Meals occurred at 0730 to 0800, 1130 to noon and 1930 to 2000, the latter time allowing section 2 (8 to 12 watches) to sleep later before the 2000 to midnight watch. Many section 1 men (12 to 4 watches) also snacked before 1600, and many section 3 men (12 to 4 watches) just before midnight.

There was not a single hour of the day when there was no one asleep. However, most section 1 men were asleep from midnight to 0300, most section 2 men from 0400 to 0700 and most section 3 men from 0800 to 1000. The mean percent of watchstanders asleep at a given time of day or night was quite similar to the percentage of non-watchstanders asleep at each hour. These patterns emphasized the human circadian predilection for nocturnal sleep.

The predominant sleeping pattern was to acquire one long sleep and a nap each day, even among non-watchstanders, and most of the crew acquired eight hours of sleep per 24-hour period. This latter observation differs from measurements made on surface ships where sleeping during the day was usually not allowed.

Nine of the crew took their oral temperature every four hours, and data from four were plotted in the report. A non-watchstander (G.P.) showed the typical circadian low point at 0400, and his degree of wakefulness was associated closely and positively with his body temperature. The three watchstanders showed circadian disturbance, with double peaks in temperature and wakefulness with maxima near the ends of their four-hour watch periods, and with the afternoon and evening maxima greater than the night and morning maxima.

Kleitman's comments about the poor quality of the racks in which the crew slept, compared to their good acceptance of the racks, led him to comment, "The absence of complaints can be explained only by the *esprit de corps* of the submarine service, the general tone of cheerfulness that prevailed on board, and the predominantly extrovert type of personality among the crew" (p. 336) [44 of the 74 crewmembers were extraverts, per a psychological survey instrument]. [Personal note: when one is sleep-deprived, the quality of the rack takes second place to its availability.]

As a result of his analysis of the *Dogfish* data, proposed the close plans, described, above, in my first chapter on research concerning watchstanding on surface vessels. He then conducted a laboratory study of several versions of close plans (157). I described this study, also, in that earlier chapter.

1949

U.S. Navy Lieutenants (jg) Robert Utterback and George Ludwig assessed a Kleitman close plan during several submarine cruises (312). The research was conducted by the Naval Medical Research Institute in Bethesda MD. The three-section, fixed close plan the Lieutenants used is shown, below (Plan 22). They referred to the plan as a 3-3-2 plan based upon the lengths of the three watch periods. Meals on the 3-3-2 plan were served at 0630 to 0730, 1230 to 1330, 1830 to 1930, and 2330 to 0030, overlapping the watch changes.

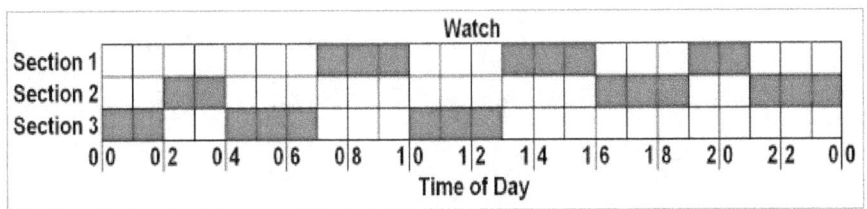

Plan 22. Underway version of Kleitman's three-section, fixed, 3-3-2, close-4 watch plan (312).

Section 1 had no unexpected fatigue issues predicted by my modeling. Section 2 did not fare as well. Predicted average cognitive performance effectiveness declined throughout the 15 days, with an average 73% effectiveness during watch periods and 89% of watch time spent BCL. By day 15 there was a 5.5-hour phase advance of the body clock with no hint of stabilization. Surprisingly, recovery took only one night of ten hours of sleep.

Section 3 fared even more poorly than section 2. Mean effectiveness was only 75% during watch periods with all watch time spent BCL. Effectiveness declined slowly across the 15 days, while there was a very small phase advance of the body clock. Recovery required two nights of ten hours of sleep. The actual measurement of body temperatures while underway (312) suggested that the FAST model may be too conservative. The model predicted no stabilization for sections 2 and 3 in 15 days, while the underway data, discussed next, suggested stabilization around day twelve (Utterback & Ludwig, op. cit., Figure 5).

The Lieutenants gathered data during two short training cruises and two longer simulated war patrols on the *USS Cubera* (SS 347), *USS Tusk* (SS 246), and *USS Sirago* (SS 485). The cruises allowed comparisons of the close plan to the classic maritime fixed plan in the following manner and with the following numbers of subjects (Figure 21, next page).

The data they gathered focused upon body temperature patterns and on preferences for one plan or the other. From their data, the Lieutenants concluded the following:

> The body temperatures on the 3-3-2 (close) plan were consistently higher than on the 4-and-8 plan. [Thus, in theory, alertness was higher on the 3-3-2.]

181

Cruise	Plan	Duration	Number of Men			
			Sect 1	Sect 2	Sect 3	Total
I	4 and 8	3.5d	6	5	6	17
II	Close-4	4d	6	4	5	16
III	4 and 8	19d	10	11	8	29
IV	Close-4	21d	8	10	10	28

Figure 21. Groups used for comparison in the 1949 submarine study (312).

Body temperatures on the 4-and-8 plan were relatively low from 0200 to 1200 even after 18 days, while they were relatively high on the 3-3-2 plan after twelve days.

Acclimatization to the 3-3-2 plan was more rapid and more complete than to the 4-and-8 plan. [Thus, in theory, the men felt less *malaise* on the 3-3-2.]

The 3-3-2 plan "won the qualified approval of a majority [30 of 59] of the men." Their preferences were:

Section	For	Against	Undecided
1	15	2	2
2	11	7	1
3	4	14	3
Total	30	23	6

Figure 22. Preferences about the close 4 plan (312).

Section 3 was not wild about the new plan. Though the close plan did appear to have physiological advantages in terms of the acclimatization and stabilization of the circadian rhythm in body temperature, it was certainly not an unqualified success in terms of acceptance.

There seems to be a 20-year gap here in published research literature concerning watchstanding in submarine operations.

1960s

During the 1960s, U.S. Navy submarines began operating on an 18-hour work-rest cycle with six-hour watches. I have been curious about why this decision was made.

Citing the 1939 version of Kleitman's extensive review, *Sleep and Wakefulness,* Kleitman and Jackson in their work for the Naval Medical Research Institute had noted that the circadian rhythm of body temperature was not inherent and could be "shifted, inverted, shortened, or lengthened, by appropriately modifying one's routine of work, play, meals, and sleep" (157). We know now that this is not an accurate statement. Later, Kleitman dedicated chapter 18 of the update to his comprehensive, definitive review of the literature to the modifiability of the circadian rhythm (155). There was evidence for the extensive modifiability of some rhythms of the body, though not body temperature. Kleitman noted that the circadian rhythm in sleep and wakefulness was due to "individually acquired mechanisms" in cerebrocortical and endocrine functions.

Thus, it seems that the prevailing, science-based opinion in the 1960s was that crewmembers could and would acclimatize, physiologically, to an 18-hour "day." Subsequent applied research has indicated that this opinion was wrong, though it was not an inappropriate conclusion based upon data available at the time.

1969

LT William Stolgitis theorized in his master's thesis at the Naval Postgraduate School that the 6-and-12 watch plan would encourage better performance among crewmembers (299). Stolgitis based his reasoning about the 6-and-12 watch plan upon two correct assumptions. The first assumption was that the work demand of a six-hour watch would not be significantly different from that of a four-hour watch. The second assumption was that the twelve-hour inter-watch break would allow for better quantity and quality of sleep than an 8-hour break. More about this theory, below.

The three-section, 6-and-12 plan recommended by Stolgitis is shown, below (Plan 23). All watchstanders stood watch for six hours and were then off for 12 hours. Then they repeated the cycle. Stolgitis also specified ship's work times for each day within the 0600 to 1800 period. On the 18-hour work-rest cycle, watchstanders slept about seven hours per 24 hours (T. L. Kelly, Grill, Hunt, & Neri D F, 1996; T. Kelly, Ryman, & Pattison, 1996; T. Kelly, Grill, and colleagues, 1996).

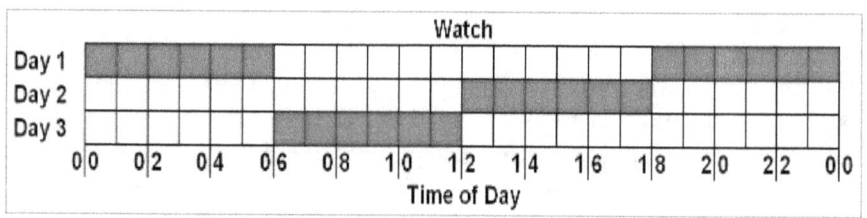

Plan 23. Three-section, fixed, 6-and-12 watch plan (299).

Writing about Stolgitis' thesis a few years later, Verne Johnson and Paul Naitoh of the Navy's personnel research group at Point Loma, San Diego, California, noted the following (145).

> Stolgitis established that the average time spent on daily watch and work duties was 12.33 hours for the 4/8 schedule, and 11.67 hours for the 6/12 schedule; thus, the two schedules produced almost equal work output. On the average, 5.82 hours of sleep were available out of the 8.67 hours for rest-recreation under the 4/8 schedule, while 8.66 hours for sleep were available out of the 9.67 hours for rest-recreation under the 6/12 schedule. By dividing the average potential daily sleep periods in hours by the average daily rest and recreation periods in hours, Stolgitis obtained an index of Sleep Cycle Efficiency (SCE)... For the 4/8 schedule, Stolgitis found an SCE of 0.67. In other words, 67% of the daily rest-recreation period was used for sleeping. A higher SCE of 0.89 was found for the 6/12 schedule... [C]rews in some nuclear submarines preferred the 6/12 schedule to the traditional 4/8 cycle as they found the 6/12 schedule more comfortable. A particularly desirable feature of the 6/12 schedule was that once in every three nights, the crew[member] has a chance to get an uninterrupted stretch of free time of approximately 10 hours 30 minutes; time enough for long uninterrupted sleep if desired. (p. 24)

I wrote this comment in 2003 (207):

> Stolgitis' arithmetic comparison of the dogged 4-h watch schedule and the [6-and-12 plan] used here did not use reported sleep times. If it had, it was likely that sharp increases in reported sleep times during the third-night, nocturnal sleep periods would have been noted, compared to sleep times reported for the other two sleep periods. This kind of pattern was found for US Coast

> Guard cutter crewmembers working on the traditional maritime 1-and-3 (4/8) work-rest cycle (213). Watchstanding ceased temporarily about every 10 days when the cutter tied up at a dock. Recovery sleep peaked during these periods that did not require watchstanding. (p. 5)

Reading again through Stolgitis' thesis, I am struck by his lack of reference to the circadian rhythm. Stolgitis did not discuss the likelihood that humans could or could not acclimatize to an 18-hour work-rest cycle. That lack suggests that Kleitman's opinion of circadian modifiability still held sway in 1969. However, my modeling of the 6-and-12 plan indicated an average cognitive performance effectiveness for watch periods of about 89%, with about 35% of that time spent BCL. The prediction for the circadian rhythm was relative stability, with a three-day oscillation of about +/- half an hour, and just one ten-hour night of sleep needed for recovery. In the context of the other modeling results reported in this book, these are pretty good numbers! Perhaps this predicted effect upon cognitive performance is close to reality and thus is part of the reason that the 6-and-12 plan has persisted in U.S. submarine operations.

Though I am unable to confirm my suspicions, I suspect that the influences of Kleitman's work for the U.S. Navy medical research community and Stogitis' thesis played important roles in the decision by the U.S. Navy to adopt and then hold to the 18-hour work-rest cycle in nuclear submarines. Looking back from 1981, twelve years after the publication of Stolgitis' thesis, Arthur Beare, Robert Biersner, Kenneth Bondi, and Paul Naitoh of the U.S. Naval Medical R&D Command in a report written for the Naval Submarine Research Laboratory cited that thesis and then noted, "Informal experiments with various watch schedules [watch plans] gradually led to the adoption of a 6-hour watch period for submarines" (19). Perhaps a reader of my book will shed further light on the process that led to the use of the 18-hour day in U.S. nuclear submarine operations.

Subsequent applied research has indicted the 6-and-12 plan. The U.S. Navy investigators mentioned just above found that the 18-hour work-rest cycle, coupled with the absence of strong photic *Zeitgebers* (daylight-darkness time cues) caused the circadian rhythms of the watchstanders to desynchronize from the 24-hour daily cycle (225). Tamsin Kelly, Jeffrey Grill, Phillip Hunt, and Dave Neri of the U.S. Naval Health Research Center found that, in some cases, the submariners' circadian temperature rhythm would free run with a period of about 24.5 hours instead of entraining to the 24-hour clock (225). Also, Karl Schaefer, David Kerr, Arnold Buss, and Erhard Haus found that some watchstanders may have developed 18-hour cycles in addition to 24-hour and other cycles (279). These studies are discussed in more detail, below.

The inability of nearly all watchstanders' circadian rhythms to entrain to an 18-hour work-rest cycle was not surprising: the limits of entrainment were shown in subsequent years to be about a 23- to 27-hour cycle in dim light (320, 322). Also, "[I]t seems probable that these findings reflect a masking effect of the sleep-wake cycle" (63).

Meanwhile, also in 1969 but on a different subject, Robert Wilkinson and R.S. Edwards in the U.K. were funded by the Royal Naval Personnel Research Committee of the Medical Research Council to conduct a shore-based investigation in which they compared a two-section plan to a three-section plan (323). The two-section plan used watches of seven- and five-hour lengths, a mealtime-oriented modification of the 6-and-6 plan. Seventeen enlisted men served as subjects across a twelve-day experiment. For the two-section plan, six men stood the five-hour 0700 to noon and 1700 to midnight watches, and five men stood the seven-hour midnight to 0700 and noon to 1700 watches. This was a variation on Plan 5, above. For the three-section plan, six men stood the classic dogged maritime plan (Plan 4).

In the longer, seven-hour watch periods of the two-section plan, the subjects worked on varied cognitive tasks during the watch. In the shorter, five-hour watch periods of the three-section plan, the subjects worked on only one cognitive task throughout the watch. I did not try to model the two-section watch plan. Obviously, the midnight to 0700 watch would cause problems.

The investigators' main interest was in comparing a circadian-stabilized plan (two-section) to a rotating plan (three-section) with the hypotheses that the non-rotating plan would stabilize the body clock and also provide better cognitive performance. The investigators reported data from the first and last three days of the experiment.

Though the cognitive performance results were mixed, "on balance, the two-man system [two-section, stabilized plan] was superior to that using three men [three-section, rotating plan]." The amplitude of the circadian rhythm in body temperature became flattened and slightly inverted for the two-section crew that stood the midnight to 0700 watch.

The investigators noted that "The argument for stabilizing working hours rests mainly on the claim that those people who have to handle the night shift will be able, gradually, to adapt to working and sleeping at unusual times, provided they do so continuously without returning for odd days to the more normal routine. That this happened in the two-man/night routine is suggested by the partial change in the circadian body temperature curve from the first (low level at night) to the last three days (moderately high level at night."

They continued, "[I]t follows that by the last three days the two-man system was better attuned to night work than the three-man one in which no such adaptation of body temperature was seen. ... [Thus,] we should expect the maintenance of performance level from first to last three days to be relatively better in the two-man than in the three-man system. This was the case with the Decision-taking but not with Vigilance or Adding."

However, it appeared that for Vigilance and perhaps Adding it was the changing of tasks within the watch period that caused the superiority of the two-section plan, while Decision-making was impaired by the changing of tasks. The investigators noted the limited applicability of these kinds of shore-based experimental results to at-sea operations due to differences in the timing of sleep.

1977

According to Emile Caille and J.L. Bassano of the Centre d' Études et de Recherches de Psychologie Appliquée (CERPA) in Toulon, France, the French Navy used a variable, "anhemeral" (non-circadian), 1 in 3, rotating watch plan (Plan 24) with a 72-hour cycle and watches of four, five and three hours (39). The investigators were able to compare this plan with a fixed, "hemeral" (circadian) plan (Plan 25) during the simulation of a 30-day submarine mission with a volunteer crew of 24 men. The two plans are depicted, below. The idea of the "hemeral," or circadian, plan was to fix a period of eight hours of sleep at a specific time of day. Submariners' body clocks would then adjust accordingly. Thus, the change from the normal day-work, night-sleep pattern to the hemeral plan was only a phase shift. They described the anhemeral, or non-circadian, plan as introducing non-linear changes in the work-rest cycle frequency, compared to a normal day-work, night-sleep pattern.

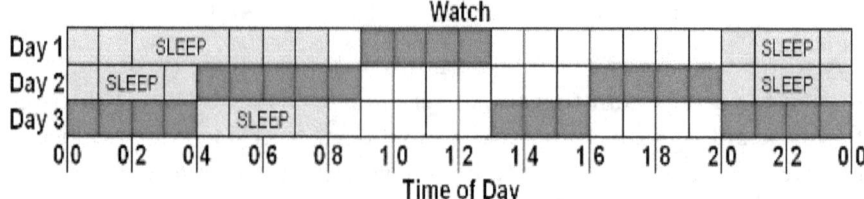

Plan 24. Three-section, counterclockwise-rotating, 4-5-4-4-3-4, French Navy traditional watch plan with fixed sleep periods (39). "Anhemeral" (non-circadian). Note the "dogging" method of extending one watch from four to five hours and the shortening of another to three hours to cause rotation. Note also the generally counter-clockwise rotation of watches from one day to the next.

My modeling of Plan 24 predicted an average cognitive effectiveness during watch periods of about 85% with about 78% of watch time spent BCL. The body clock remained stable, and two nights of ten hours of sleep were needed for recovery. For my modeling of Plan 25, see the results for the plan studied by Colquhoun and colleagues (Plan 8). The predictions for section 3 were worse than those for Plan 24.

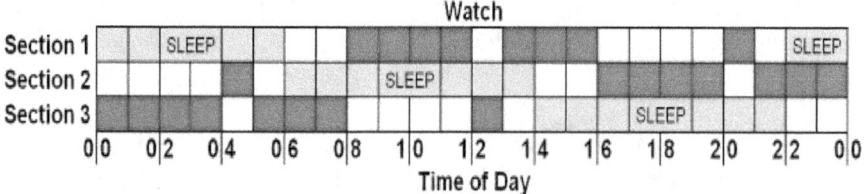

Plan 25. Three-section, fixed, 8-and-16 watch plan with fixed meal times and sleep periods (39). Experimental "hemeral" (circadian). Note the three eight-hours sleep periods and the one-hour watches at meal times. This one-hour watch change probably would not work well in operations. Without these one-hour breaks for meals, this plan simply becomes a 1 in 3 plan with eight-hour watches starting at midnight, 0800 and 1600.

These investigators recorded the electroencephalogram (EEG) during sleep periods and during watches, and critical flicker fusion threshold (CFF) as an index of cortical arousal in the brain. They sampled circulating hormone and ion levels in the blood. They also tested hand tremor, visual pursuit tracking paired with auditory reaction time (dual task), short-term memory, longer-term memory, and form and color coding. They acquired mood ratings for seven dimensions of mood.

The following measures displayed statistically significant enhancement in the hemeral plan compared to the anhemeral plan: short-term memory, code learning, dual task (both tasks) performance, tremor, mood, CFF, and cortical desynchrony (greater activation of the brain's cortex). The process of falling asleep was less distorted in the hemeral plan than in the anhemeral plan. Entrainment of biochemical circadian rhythms was superior in the hemeral plan. The investigators concluded, "The strong advantage of the second [hemeral] alternative compared to the first [anhemeral] is evidenced in the sleep process, behavioral efficiency, mood, and circadian biochemical parameters." This study showed the advantages of using a fixed, guarded sleep period, and the possible advantages of using a 1 in 3 plan with eight-hour watches.

1978

Peter Colquhoun of the PCPU, M. Paine of the historic Haslar Royal Naval Hospital and Ann Fort of the University of Manchester had the opportunity to collect hourly oral temperatures during two 48-day cruises in a study sponsored by the Royal Naval Personnel Research Committee (58,59). Their aim was to describe the effects of a rapidly-rotating shiftwork plan on the circadian rhythm in body temperature. The investigators observed that, in the submarine,

> [T]he natural diurnal *Zeitgebers*, such as changes in daylight and ambient temperature are, of course, absent. However, a daily pattern of artificial *Zeitgebers* exists; for example, the lighting is reduced during "normal" sleeping hours, and the times for meals, "clean-ship," and recreational activities are typically the same as those on surface vessels. (1978)

This is a somewhat different, more structured pattern of daily activities than observed by Kleitman in 1949 on the *Dogfish*. The latter pattern seems less conducive to acquiring adequate sleep. I wonder if lighting and work scheduling practices changed for the worse, with respect to sleep hygiene, in U.S. submarines between 1949 and 1978?

The rapidly rotating watch plan observed in this study was either Plan 6 or Plan 7, above, or maybe one cruise was spent on each plan: the investigators' 1978 paper shows Plan 7 (their Figure 1), while their 1979 paper shows Plan 6 (their Table I). The investigators labeled this a "traditional" plan, implying its regular use in the Royal Navy. Eight sonar ratings followed this plan and provided temperature data. The investigators compared this plan to a fixed watch plan that, most likely, was Plan 1, above. Four officers followed this fixed plan and provided temperature data.

The rotating plan was associated with a gradual reduction in circadian rhythm amplitude and disintegration of crewmembers' rhythms. For fixed day work, the circadian rhythm was not affected. For fixed night work, the rhythm entrained to the new work-rest schedule in about a week, by phase delay. This observation supported the idea that the submarine environment may be conducive to fixed night watches. Two officers who split their sleep displayed bimodal circadian rhythms, much as observed by Utterback and Ludwig 1n 1949, above.

On the basis of their observations, the investigators recommended using a 1 in 3, eight-hour (8-and-16), fixed work plan in submarines (57, 58). They expected that the fixed periods would allow rapid re-alignment of circadian rhythms to the new work-rest schedule in the first week of the cruise, and that the 16-hour off period would prevent sleep fragmentation and restriction. This recommendation agrees for the most part with that made by Caille and Bassano in 1977, above, and also as a result of a 2010 symposium on submarine watchstanding plans, discussed below (249). Alternatively, they recommended a modification of the Close plan suggested by Kleitman in 1949 in which the eight hours of watch are completed within a fixed, twelve-hour period (close work). Thus, 3 on, 2 off, 3 on, 2 off, 2 on, 12 off, or similar.

1979

As I noted, just above, the 18-hour work-rest cycle, coupled with the absence of strong photic *Zeitgebers* (daylight-darkness time cues), may have caused some watchstanders to develop 18-hour cycles in addition to 24-hour and other cycles (279). Karl Schaefer, David Kerr, Arnold Buss, and Erhard Haus, working for the U.S. Naval Submarine Research Laboratory in Groton, Connecticut, collected non-invasive physiological data from submariners during patrols. The investigators' objective was to study further the disintegration of the circadian rhythm in body temperature observed by Colquhoun and colleagues (58).

The investigators collected control data from five subjects during a "stressful" pre-patrol period. These subjects then stood the 6-and-12 watch plan while underway for about 7 1/2 weeks. Non-invasive physiological measurements were made during the first two weeks and again during the last week and a half of the patrol. The investigators applied Halberg's least-squares cosinor analysis to the data, at frequencies with periods from ten to 48 hours (121). Using a test statistic for which the null hypothesis was a circadian rhythm amplitude of zero, and with a statistical confidence level set to 95%, they were unable to detect statistically significant circadian rhythms in body temperature, pulse rate, respiratory rate, or blood pressure during the control period or during the two data collection periods of the patrol (their Table 1).

The investigators showed a graphic frequency distribution over time for the rhythms for the five subjects (their Figure 1). There were clusterings near the twelve-hour, 18-hour, and circadian (24-hour) cycle lengths. This descriptive assessment indicated an overall loss of 24-hour rhythms during the patrol with a concurrent gain of 12- and 18-hour rhythms. Since the investigators did not specify a criterion for detecting activity at a given frequency, and reported being unable to detect statistically significant amplitudes for circadian rhythms, it is difficult to state reliably that 18-hour rhythms appeared.

The investigators then collected data from three more subjects on the 6- and-12 plan and three subjects on a "regular 24-h schedule" (day work and night sleep, I assume) in each week during a second seven-week patrol. The investigators presented a tabular frequency distribution over time for the rhythms for the two sets of three subjects. They noted clusters around 12-, 18-, 24-, 36-, 42-, and 48-hour cycles (their Table 3). This descriptive assessment indicated an overall loss of 24-hour rhythms during the patrol with a concurrent gain of 12- and 18-hour rhythms, and also 36-hour rhythms. However, once again the investigators did not specify a criterion for detecting activity at a given frequency. Thus, again, it is difficult to state reliably that 18-hour rhythms appeared.

Though the results of the data collection and analysis may seem suggestive of the appearance of an 18-hour rhythm in the watchstanders on the 6- and-12 plan, I am a bit skeptical. My own work with cosinor analysis and Fourier transform has indicated to me that the least-squares approach to frequency analysis is very "noisy." That is, it tends to generate false positive findings of frequency content in a signal. If Schaefer and colleagues had reported the appearance of 18-hour rhythms that were statistically significant at the 95% or even 90% level of confidence, I would have been convinced that real 18-hour cycles may well have appeared. However, they did not report any levels of confidence for their findings of 18-hour cycles.

Instead, I believe that work-rest demands before and during the patrols suppressed the amplitude of the dominant circadian rhythm, leaving a noisy, mixed spectrum of low amplitude responses to circadian, circasemidian, and 18-hour influences and their harmonics. While these responses are of some interest, they provide little or no direct support for the development of an 18-hour body clock that may have replaced the circadian rhythm. However, see the work of Naitoh et al. in 1983, below (225), which does provide direct support for an 18-hour sleep cycle, but not an 18-hour oral temperature cycle.

1981

Arthur Beare, Robert Biersner, Kenneth Bondi, and Paul Naitoh of the U.S. Naval Medical R&D Command, working for the Naval Submarine Medical Research Laboratory at Groton, Connecticut (NSMRL), used daily logs to examine hours of work and sleep in Fleet Ballistic Missile submarines (19). Forty-six enlisted men provided data; thirty of these stood the 6-and-12 watch plan.

These investigators noted that about 75% of the enlisted men in the submarine fleet were on the three-section 6-and-12 plan, while most officers and senior enlisted used a four-section, 6-and-18 plan. I have not described a four-section 6-and-18 plan here previously. You could organize this plan with a start time around midnight (Plan 26, below), or around 0300 (Plan 27, below).

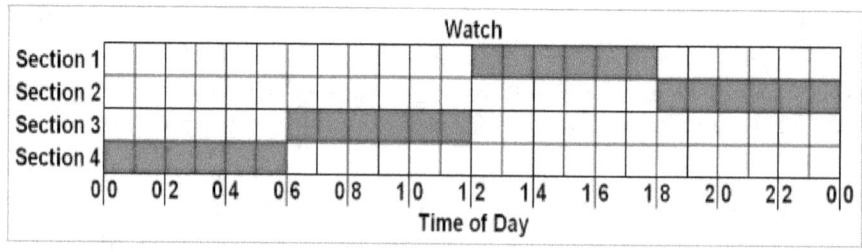

Plan 26. Four-section, fixed, 6-and-18 watch plan starting at midnight.

My modeling for section 1 and section 3 in Plan 26 was not necessary, since they should have no unexpected fatigue issues. For section 2, working evenings, the predicted average cognitive performance was 96% during watch periods, with 7% of that time spent BCL. There will be a phase delay of about 1 hour through day 9, then circadian stability, and one night of ten hours of sleep is required for recovery. However, these relatively good numbers apply if and only if late sleeping to 0800 is allowed. Sleeping to only 0700 causes surprisingly severe fatigue issues at the end of each evening watch period.

My modeling for section 4, working the midnight to 0600 watch in Plan 26, predicted an average cognitive performance of 72% during watch periods, with 100% of that time spent BCL. There will be a slow phase delay of two hours that will not have stabilized at 14 days. More than three nights of ten hours of sleep will be needed for recovery.

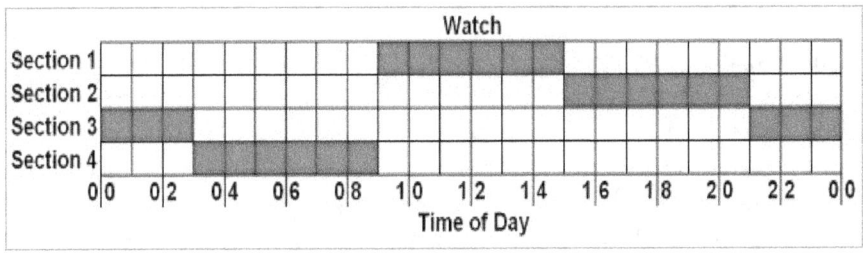

Plan 27. Four-section, fixed, 6-and-18 watch plan starting at 0300.

My modeling for section 1 and section 2 in Plan 27 was not necessary, since they should have no unexpected fatigue issues. For section 3, working across midnight, the predicted average cognitive performance was 87% during watch periods, with 56% of that time spent BCL. There will be a phase delay of about two hours through day 14 with no stabilization. More than three nights of ten hour of sleep will be required for recovery.

For section 4, working the 0300-0900 watch, the fatigue model was quite interesting. First, the predicted average cognitive performance was 91% during watch periods, with 38% of that time spent BCL. There will be a phase advance of about three hours through day 8 followed by circadian stabilization. No nights will be required for recovery. The interesting aspect of this modeling result was that the unacceptable levels of fatigue occurred only in the first six of the 14 modeled watch periods. Thus, even with strong photic *Zeitgebers* (sunlight), this section will have a fatigue-related a safety issue for only about the first week of a cruise.

Back to the data. These investigators were following up on the observations reported in 1979 by Schaefer and colleagues, above (279). Among the 46 subjects who provided usable data, 23 were aboard the first submarine during a 40-day deployment. Of these 23, only seven stood the 6-and-12 watch plan. On the second submarine, which was on a 70-day patrol, 19 of the 23 subjects stood the 6-and-12 watch plan.

Also, two subjective rating scales were added to the diary during the second submarine's patrol: the Stanford Sleepiness Scale (SSS) (132), and four sleep quality questions from Hartman and Cantrell (126, 201). The two subjective ratings of most interest from the latter were SQ2, "How rested do you feel?" which was rated on a 1 ("well") to 4 ("not at all") scale, and SQ3, "Do you feel you could have used more sleep?" which was rated as yes or no.

The investigators analyzed a 31-day block of data from the first submarine, and they concatenated three ten-day blocks from the second submarine's patrol into a 30-day block for analysis. Watchstanders on the 6-and-12 plan stood watch for an average of 8.1 hours per 24 hours, while those on the 6-and-18 plan stood watch for an average of 6.2 hours. The across-subjects variability was small for the amount of time spent on watch. Non-watch additional duties accounted for 2.6 hours and 3.7 hours, respectively, but with a great amount of across-subjects variability. Time spent in technical studies, which is emphasized on nuclear submarines, accounted for one or two hours per day. The total workload across all subjects, the two watch plans and the two submarines was "about 12 hours out of every 24, or 84 hours a week."

The investigators tallied the numbers of hours sleep per 24 hours, the within-subjects variability in these amounts experienced by individuals across the days of data collection, and the number of episodes of sleep (split sleep) each 24 hours. From the second patrol, they also assessed the SQ2, SQ3 and SSS responses.

On the first patrol, all of the subjects acquired their eight hours of sleep per 24 hours varying about +/-1.25 hours, and splitting their sleep. On the second patrol they acquired about 7.5 hours per 24 hours with about the same amount of variability and split sleep.

On this second patrol, the subjects rated their sleepiness in the range of 2 to 4 on the SSS, *i.e.*, "Functioning at a high level, but not at peak; able to concentrate," through "A little foggy; not at peak; let down." Ratings of 1 on the scale, *i.e.*, "Feeling active and vital; alert; wide awake," were relatively rare, as were ratings of greater sleepiness in the range of 5 to 7. It is this latter range that usually invokes concern about the ability to meet complex work demands safely. Additionally on the second patrol, they rated their level of being rested (SQ2) as about a 2 (not quite well rested) and said that they could have used more sleep (SQ3) about half of the time.

The investigators reported other aspects of sleep regularity and some mood ratings, and provided an excellent discussion of their findings. In the end, they concluded that the workload, in terms of hours spent on watch, on ship's work and studying, was "substantial." The sleep of the watchstanders was seen to be normal in terms of quantity, but somewhat fragmented and of slightly less quality than desired. However, they did not see evidence that sleep deprivation was a significant stressor. They noted that mood scores for "Activity" and "Happiness" were not as high as desired and suggested more study of this finding.

1983

As I noted above, the 18-hour work-rest cycle, coupled with the absence of strong photic *Zeitgebers* (daylight-darkness time cues) caused the circadian rhythms of the watchstanders to desynchronize from the 24-h daily cycle (225). Paul Naitoh, Arthur Beare, Robert Biersner, and Carl Englund, all affiliated with U.S. Navy research groups, collected data on sleep and activity, body temperature and mood on 15 submariners during the first (after three days underway), middle and last weeks of a ten-week patrol conducted by a fleet ballistic missile nuclear submarine. This study was characterized as a follow-up to the 1979 study by Schaefer and colleagues (279), in which the 18-hour watch cycle and 24-hour social and meal cycle were in conflict.

The investigators quantified the period length of the sleep-wake cycle with Halberg's cosinor analysis. Presumably, they used the range of period lengths, ten to 48 hours, as did Schaefer and colleagues. Their results are shown in Figure 2.

Week	Sample	Primary Cycle	Secondary Cycle
First	n = 11	18.5 hours, r^2 = 68%	25.3 hours, r^2 = 25%, n = 6
Middle	n = 8	18.7 hours, r^2 = 67%	25 hours, r^2 = 27%, n = 4
Last	n = 7	19.1 hours, r^2 = 73%	25.75 hours, n = 1 10.5 hours, n = 1

Figure 23. Submariners' circadian cycle lengths (225).

The "sample" was the number of analyzable reports for the week among the data provided by the 15 subjects. The primary cycle column data showed that there was a strong relationship between the 18-hour watch plan and the sleep-wake cycle. The r-squared percentage shows how well a perfect cosine curve represented the variable data collected that week on that individual. An average r-squared of 70% is quite good. These primary cycle data assure us that the subjects were sleeping pretty much in harmony with the watch plan's 18-hour cycle length. Each week, a few folks also had a secondary peak in the sleep-wake cosinor results. The secondary peak, approximating the 24-hour day in its period length, was weak (low r-squared) and was almost absent in the last week of the patrol. These results suggested that the tendency to sleep on a 24-hour cycle died off as the patrol progressed.

The investigators analyzed group mean cosinor results acquired from oral temperature measurements. They noted that the mean amplitude of the circadian rhythm in oral temperature was relatively low during all three weeks of data collection, but did not change greatly across the patrol. However, the phase relationship of the circadian rhythm with respect to midnight became much more variable. Usually, the timing of the high point (acrophase) of the circadian rhythm in body temperature occurs in the evening, maybe 20 or 22 hours after midnight. In the first week of the patrol, the subjects were relatively well synchronized, with a circadian rhythm acrophase at about 2100. During the patrol, synchrony for the circadian rhythm was lost, and no significant synchrony appeared for an 18-hour rhythm. Their analysis of group mean data led the investigators to conclude that there was "no 18-h rhythm in oral temperature during any of the three phases of this study" for group means.

The investigators also assessed each subject's oral temperature data for the strongest rhythm (highest r-squared) across the range of rhythms. They had complete data for seven subjects for this analysis. In the first week, three subjects' strongest cycle lengths were 24 and 25 hours, two were 17 and 18 hours, one was 32 hours and one was 36 hours. In the second week, two of the 24- and 25-hour subjects stayed the same while one moved to 48 hours, the 17- and 18-hour subjects moved to 24 hours, the 36-hour subject moved to 18 hours, and the 32-hour subject moved to 16 hours. In the final week of the patrol, one of the original 24-hour subjects moved 6 hours, the 48-hour subject from week 2 moved to 44 hours, and the remaining five subjects were at 24, 25 and 26 hours. The r-squared values for all of these cosine curve fits were relatively small, from 14% to 43%, increasing slightly across weeks from a mean of 21% in the first week to a mean of 29% in the third week. This individualized approach to the data led the investigators to conclude, "[T]he 18-h watch-standing schedule did not significantly alter the basic period of the oral temperature rhythm."

While the 18-hour work-rest cycle entrained the sleep-wake cycle of the subjects in this study, as it had in other, non-maritime studies, its effects on the circadian rhythm of body temperature and on mood (not described here) were weak. The investigators noted the lack of synchrony in this patrol between the sleep-wake cycle, entrained to an 18-hour day, and the body temperature cycle, continuing its attempt to operate on a 24-hour cycle. They suggested, "a spontaneous internal desynchronization (321) may be an unavoidable consequence of life under the 18-h watchstanding schedule." However, they were unsure what the effects of this desynchrony upon mental performance and mood might be.

I add here that the occurrence of some 18-hour periodicity in the sleep-wake cycle was forced upon the sections by their watchstanding plan. Thus, though the plan forced some 18-hour periodicity in the sleep function, the body clock (suprachiasmatic nucleus) resisted the change, leading to desynchrony between the two periodicities. Generally, this kind of situation leads to a feeling of malaise.

1989

As reported by Donderi and colleagues in 1995 (96, p. 2), R.J. Strong and D.C. Brown of the Institute for Naval Medicine (INM) had assessed eight control room personnel, four in each section, over 14 days while underway on a Royal Navy submarine (300). The port section stood watch from 0700 to 1300 and 1900 to 0100, while the starboard section stood watch from 0100 to 0700 and 1300 to 1900. The subjects accomplished response time and visual search tasks at 0600, 1100, 1800, and 2300. Many data points were missing, but it appeared that the port watch (the "day" watch) suffered a greater decline in performance than the starboard watch. This result was somewhat counterintuitive and may have been caused by the small sample size.

1996

As I noted above, watchstanders slept about seven hours per 24 hours on the 18-hour work-rest cycle (150). Tamsin Kelly and colleagues, working for the Naval Personnel Research Center in San Diego, cited the 1983 study by Naitoh and colleagues (225, above) and were curious whether the circadian rhythm might actually free run under the 6-and-12 watch plan. This study was conducted during the second, fourth and sixth weeks of a patrol by a fleet ballistic missile submarine, the *USS Georgia* (SSBN 729).

Tamsin Kelly and Dave Neri of the U.S. Naval Health Research Center, Jeffrey Gill, and Phillip Hunt acquired salivary melatonin samples, sleep logs, wrist activity (WAM) data, and performance assessment battery (PAB) data that included mood ratings that were administered on hand-held computers. Salivary melatonin had been shown to be an excellent marker for the presence of a normal circadian rhythm (172). The sleep log and WAM data indicated, "[T]he subjects averaged a little less than 7 hr of sleep out of every 24 hr, occurring during an average of 9.5 sleep periods per week. The average duration of an individual sleep episode was about 5.5 hr. Subjects averaged 65 hr on duty per week (including supplementary duties)." Task performance and mood were reasonably constant across the three weeks. The investigators concluded, "Subjects appear to get sufficient sleep and to maintain acceptable performance levels on this work schedule" (150). This was a preliminary report about the data collected during the patrol.

1999

Subsequently, Tamsin Kelly, Dave Neri, Jeffrey Grill, David Ryman, Phillip Hunt, Derk-Jan Dijk, Theresa Shanahan, and Charles Czeisler reported the results of the salivary melatonin analyses from the patrol on the *USS Georgia* (151, abstract only). They noted that the 18-hour day imposed by the 6-and-12 plan "is too short for human circadian synchronization, especially given that there is no bright-light exposure aboard submarines. However, crew members are exposed to 24-h stimuli that could mediate synchronization, such as clocks and social contacts with personnel who are living on a 24-h schedule." Data from 20 crew members working the 6-and 12 indicated a free-running circadian rhythm with an average period of 24.35 hours. They concluded, "These data indicate that social contacts and knowledge of clock time are insufficient for entrainment to a 24-h period in personnel living by an 18-h rest-activity cycle aboard a submarine."

Chapter 10. Submarine Watchstanding Studies 2000-2010

2001

LT Simonia Blassingame, in her NPS master's thesis used data from the U.S. Naval Submarine Medical Research Laboratory (NSMRL) Watchstanding Survey to help determine

> what a sub-sample of this population think about their sleep habits and ... if there are differences in the reported amount of sleep between sailors in four different operational environments: 1) at sea, 2) in port, 3) on shore duty, and 4) on leave. (22)

The NSMRL Survey was shown in Blassingame's report as Appendix F. This is the only report that I have seen in which the NSMRL Survey has been used. According to Blassingame, the survey

> was originally designed by SurgCDR Steve Ryder of the Royal Navy and was later modified by Dr. Christine Schlichting under Office of Naval Research (ONR) sponsorship. The current effort began in October 2000 under the sponsorship of ONR via CDR Steve Ahlers and also has the support of Commander Submarine Group Two (COMSUBGRU TWO). This effort originated from the ideas and initiative of SurgCDR Ryder and Dr. Schlichting. CDR Wayne Horn and LT Jeff Dyche serve as the principal investigators of this study designed to evaluate improvements on the quality of life (QOL) and performance of US Naval submariners via manipulation of the watchstanding schedule.

> The survey contains 37 questions with six demographic and submarine work-rest schedule related questions. The final two questions are free response questions pertaining to the current watchstanding routine. Demographic questions cover current duty assignment, age, rank, rating, submarine qualification, number of children under the age of five, nuclear designation, and current watch schedule. The remaining questions pertained to the habits and schedule at four conditions: 1) at sea, 2) in port, 3) on shore duty and 4) on leave. The expected responses were either a direct answer to a specific question (e.g., How many hours of

sleep, per 24 hours, do you think you need to function at your best?) or used Likert scaling of never, rarely, sometimes, often, frequently or always. (22, pp. 21-22)

These latter six descriptors of frequency are, in fact, discriminable from each other reasonably well by most people (10, p. 132).

Responses were available from 143 enlisted submariners. I have copied Blassingame's table showing their reported watch plans in Figure 24. Nearly 78% worked the 6-and-18 plan. There were some other, strange plans reported. The 6-and-8 plan and the 8-and-12 plan result in 14- and 20-hour cycles. Very strange. The 9-and-9 plan is most likely what happens when the crew is on an 18-hour cycle and there are only two qualified watchstanders for a given position. The 12-and-6 plan does not seem sustainable except by a person whom, physiologically, is what we call a short sleeper; someone whose sleep need is two or three standard deviations below the population mean of about eight hours. These people exist, but there are very few (20, 219).

Reported Schedule	Number of respondents	Percentage of respondents
6 on 6 off	3	2.1%
6 on 8 off	1	0.7%
6 on 12 off	111	77.6%
6 on 18 off	3	2.1%
8 on 12 off	1	0.7%
9 on 9 off	1	0.7%
12 on 6 off	3	2.1%
12 on 12 off	8	5.6%
14 on 10 off	3	2.1%
24 hours	1	0.7%
No response	1	0.7%
Other	7	4.9%
Totals	143	100.0%

Figure 24. Responses of 143 enlisted submariners about their experience with various watch plans (22).

Questions about sleep pertained to four conditions: at sea, in port, shore duty, and on leave. The sleep questions were:

How many hours total sleep do you usually get per 24 hours?

On average, what is the longest period (in hours) of uninterrupted sleep you get per 24 hours?

How many hours of sleep, per 24 hours, do you think you need to function at your best?

The distribution of reported total hours of sleep per 24 hours at sea was bimodal, with peaks at 6 and 5 hours (37 and 27 responses, respectively). In port, the distribution was multimodal. Shore duty sleep amount was bimodal at 8 and 7 hours, and there was a strong mode (54 responses) at 8 hours while on leave. Thus, "Submariners reported getting less sleep while 'at sea' than in other conditions."

The distribution of lengths of uninterrupted sleep when at sea was pretty much unimodal at 4 hours or less. The in-port distribution was multimodal. The shore duty distribution had a weak mode at 8 hours, and the on leave distribution had a strong mode (41 responses) at 8 hours. The distributions of desired sleep were multimodal across the conditions.

Analyses of the data led to the following observations:

> Of the four operational conditions evaluated, the 'at sea' condition was the most different from all other conditions. Submariners reported getting less sleep while 'at sea' than other conditions.

> There was a positive correlation between the amounts of sleep obtained (both total and uninterrupted) and the desired amounts of sleep needed to function in all four conditions. This led to the inference that subjects who reported needing more sleep did indeed get more sleep. When in the 'at sea' condition, this correlation was much weaker indicating that subjects had much less control over the amount of sleep they get when deployed.

In that same year, LT Sarah Chapman, a Royal Australian Navy Submarine Psychologist on the *HMAS Stirling*, reported underway measurement of the sleep-wake cycle of submariners by wrist actigraphy and diaries, and the sampling of their saliva during watches for melatonin levels (49, 50). Measurements were made with 27 crewmembers on *HMAS Collins* for a week in July 2000 during joint RAN-USN weapons exercises. These subjects stood 6-and-6 two-section watches starting at 0100. Chapman also acquired similar data from 32 crewmembers for 13 days on *HMAS Waller* in August-September 2000 during a transit. Twenty-two of these subjects stood a three-section watch while ten more in the engineering section stood a two-section watch. She reviewed submarine lighting issues and the interactions among sleep, circadian rhythms and operational schedules. Though Chapman did not report the results of the actigraphy and melatonin assays (nor have I found a subsequent report), she did report observing the following fatigue symptoms among the crewmembers: 1) slowed speech, 2) delayed response to orders, 3) greater number of malapropisms and incorrect sequencing of orders, 4) delayed repetition of orders, 5) failure to acknowledge orders, 6) increase in vacant stares, irritability and minor altercations between personnel, and 7) decreased ability to acknowledge multiple sources of information during longer periods at sea. Chapman also provided fatigue risk management guidance, which I have discussed in my FRMS chapter, below.

2003

I investigated an alternative to the 18-hour submarine watchstanding plan for the U.S. Naval Medical Sub Research Lab (NSMRL) (207). Navy psychologists LT Walter Carr and LT Jeff Dyche, and Ms. Rebecca Cardenas of NTI Inc. were major contributors. The alternative plan was the Close-6 plan that I had developed as a result of my investigation of fatigue in Coast Guard cutter crews (213, Appendix G; Plan 16, above). Recall that the rotating Close-6 plan used work compression to provide one 12-hour period and one 24-hour period of time off each 72 hours. I conducted this study in the Chronobiology and Sleep Laboratory of the Air Force Research Laboratory (CASL), an isolation facility in building 1192 at the former Brooks Air Force Base in San Antonio, Texas (201). I compared the Close-6 plan to the 6-and-12 plan that was in use in submarines (Plan 23, above), and to section 3 of the classic maritime watch plan (Plan 1, above, 12 to 4 watches).

The subjects for this study were nine male submariners, two from attack submarines and seven from ballistic missile submarines. The group spent three eight-day periods living in the CASL, one period on each plan.

A side note. The midpoint of the second period was 11 September 2001, the day of the terrorist attack on the World Trade Center. Navy and Air Force commanders decided that the study should continue, though Brooks AFB went to the highest alert status. The NSMRL Chief Petty Officer was in the CASL. He coordinated with the subjects' commanders and also assured that no subject's family members were involved in the attack. We limited sharply the amount of attack information passed to the subjects until the end of that second test period. With the base restricted that day to essential personnel, we were unable to have our contract meals delivered. Lt Lisa Thiem, assisting me with the project, lived on base. She went home, emptied her refrigerator, returned to the laboratory, and fed the subjects and research team. An un-heralded act that was above and beyond the call of duty! Lisa is now a field grade officer working in security management. We are in good hands!

Back to the watchstanding study. I tested the following hypotheses.

> That the two 24-hour work-rest cycles would produce better entrainment of circadian rhythms in physiology and performance to the 24-hour clock than would the 18-hour work-rest cycle. In fact, the participants adjusted their body clocks quickly to the fixed work-rest plan of section three of the classic maritime plan.

> That both sleep quality and sleep quantity would be worse in the 18-hour work-rest cycle than in the two 24-hour work-rest cycles. In fact, more good-quality sleep was acquired on the Close-6 than on the other two plans.

> That performance and mood would be worse on the 18-hour work-rest plan than on the two 24-hour plans. This hypothesis was not supported.

> That both sleep quality and sleep quantity would be worse on the classic maritime section three plan than on the Close-6 plan. In fact, more good-quality sleep was acquired on the Close-6 plan than on the other two plans, and the need for recovery sleep was not an issue following the Close-6 plan but was definitely an issue following the classic maritime three-section plan.

> That performance and mood would be worse on the classic maritime section three plan than on the Close-6 plan. The mood portion of this hypothesis was supported: the malaise predicted to occur as a result of circadian rhythm disorder caused by the classic maritime section three plan was detected by subjective ratings on a "Fatigue" scale. The performance portion of the hypothesis was not supported.

I recommended considering the following three-section schedules for sea trials:

The Close-6 plan

A fixed, dogged-6-hour plan

A fixed, 8-hour watch plan (8-and16)

The five chief petty officers of the whole U.S. submarine fleet, the Atlantic fleet, the Pacific fleet, NSMRL, and the submarine school all thought that the Close-6 plan would be excellent for scheduling ship's work, administrative duties and training, as well as for adequate sleep. Thus, the plan was taken to sea trials (100). The underway study is described, below.

2004-2007

With sponsorship from NSMRL, LCDR (Dr.) Chris Duplessis headed up a comparison of the Close-6 plan to the submarine 6-and-12 plan (100,). The evaluation occurred aboard the ballistic-missile submarine, *USS Henry M. Jackson* (SSBN-730 Gold). LT Christopher Osborn assisted Duplessis and wrote his NPS master's thesis based upon objective and subjective data collected while underway (243). Forty subjects provided sleep and oral temperature data, and ten of them provided salivary cortisol data for approximately two weeks on each plan while underway. The circadian amplitudes and means for cortisol did not differ significantly between plans, nor did the temperature amplitudes and means. However, cosinor analysis showed a stronger circadian rhythm in body temperature with the Close-6 plan than with the 6-and-12 plan. Conversely, the 6-and-12 plan allowed significantly more sleep (7.1 hours) than the Close-6 plan (6.3 hours). The latter finding was interesting, considering the allotment of 24 hours off in the Close-6 plan. Fifty-two percent of survey respondents preferred the 6-and-12 plan, while only 15% preferred the Close-6 plan.

The conclusion was that the Close-6 plan was not superior to the 6-and-12 plan schedule in terms of physiological or subjective measures. Additionally, the chief petty officers on this first trial patrol were unable to schedule ship's work, administrative duties and training adequately. The trial was supposed to occur in four submarine patrols: two on the east coast and two on the west cost; two in ballistic missile submarines (SSBN) and two in attack submarines (SSN). However, the sea trials were stopped after one patrol and the U.S. Navy moved on to consider fixed eight-hour shifts for submarine operations.

Osborn concluded that the submarine fleet needed to improve its watchstanding schedules, especially with regard to improved sleep hygiene. He also summarized the lessons learned as described at a post-trial working group meeting that I attended, held 4 and 5 March 2004 at Submarine Base Bangor, Washington.

> Results may have been different on a fast attack submarine.
>
> Schedules for training, drills, etc. should be worked out before sailing.
>
> Exit survey questions could have been worded better.
>
> Crew education should include fatigue management.
>
> Task analysis and fatigue risk analysis should be undertaken for various watch stations.
>
> Tests should be conducted on a fixed, eight-hour watch plan and other six-hour watch plans.

In the fiscal year 2004 overview of U.S. Naval Submarine Medical Research Laboratory (NSMRL) work, Jerry Lamb, Maria Fitzgerald and Heather Huebner noted the following:

> NSMRL researchers and crew set out on the SSBN cruise following the 6/12 schedule for 3 days, then shifted to the alternative schedule. However, that schedule was first initiated at midnight, vice noon, as designed. The crew followed that schedule for 13 days, then shifted so the alternative schedule would begin at noon; this was followed for 6 days. After that, the 6/12 schedule was re-adopted for the remainder of the cruise, i.e., 11 days.
>
> The Henry M. Jackson crewmembers provided feedback on the compressed work schedule. The negative sentiments of the crew reflected how the schedule did not provide the time needed to accomplish ancillary tasks (e.g., training, qualifications, drills, and rest) between watches. When the subjects reached their 24-hour break on the 3rd day, they routinely spent at least 12 hours sleeping.
>
> A videoteleconference (VTC) was held at Naval Submarine School, Bangor, WA, on 4-5 Mar 2004. This VTC provided the researchers with frank and open communication with the crew of the Henry M. Jackson, along with representatives from Norfolk and Pearl Harbor. Various alternative schedules were discussed:

1. a fixed, 8-hour,

2. a compensated 6-hour dogged, and

3. a fixed, 6-hour dogged.

After evaluating each candidate schedule based on the Fatigue Avoidance Scheduling Tool (FAST), the group agreed that the fixed 8 shift appeared to provide the greatest promise of simultaneously entraining submariners on a biologically-sensible 24 hour "day," provide sufficient time off each 24 hour period to accommodate sleep, and allow entrainment of the appropriate circadian rhythm for watchstanders on the swing and midnight shifts, enhancing their performance levels.

NSMRL secured a no-funds extension for BUMED appropriation 03-1319, providing us the funds needed to conduct the laboratory-based evaluation, and we are currently writing a budget to secure the necessary equipment, personnel, and cover overhead charges.

Subsequently, in their fiscal year 2005 overview of the U.S. Naval Submarine Medical Research Laboratory (NSMRL) work CAPT Christopher Daniel and Jerry Lamb noted the following:

> In the unforgiving undersea environment, 24/7 operations require a rested and alert crew. Normally, humans have a daily cycle of wakefulness and sleep, the Circadian Rhythm (CR), which is driven by the sun's passage. Submerged Sailors have no daily light clues to stabilize their CR. The current watch cycle of 6 hours on watch and 12 off often leads to a destabilized, free running CR, and the possibility of standing watch at a low point in the sleep/wakefulness cycle. Because the day is only 18 hours long, the CR pattern is constantly shifting, causing further loss of alertness—the equivalent of flying eastward through six time zones every 18 hours. NSMRL has been studying how new watch schedules that more closely follow a normal 24-hour day might work. Any potential change must not only help with the CR patterns for increased alertness, but must also accommodate all of the boat's operational requirements. A recent sea trial of an 8/16 schedule was conducted on *USS Maryland* (SSBN 738) with behavioral, physiological, and psychological measurements. While the data are still being analyzed, initial indications suggest that it

improved overall alertness. As important perhaps was the crewmembers' reaction; they thought that it was much better—and that it didn't adversely impact their normal routine, operations, or drills. (87, pp. 6-7.)

In the fiscal 2005 year report for NSMRL, that study was described as follows:

> Work Unit #50513
>
> Title: Evaluating a fixed 8-hour submarine watchstanding schedule
>
> Principal Investigator: L. J. Crepeau, LCDR, USN
>
> Accomplishments (FY05):
>
> An underway study was conducted aboard the *USS Maryland* (SSBN 738) 7 May - 3 June 2005. During that time, submariners followed the FIXED 8 schedule for two weeks, then followed the standard 6/12 submariner schedule for 11 days. Two subjects served for the polysomnographic sleep portion of the study, and six sleep records were collected from each subject in each condition. These have been completely analyzed. Saliva samples were collected every 2-3 hours from 24 subjects during the entire 25 days. Samples were rapidly frozen and stored in the submarine's freezer until reaching port, then sent to the Naval Institute of Dental and Biomedical Research, Great Lakes IL.
>
> Subjects also completed PDA-administered cognitive tests and, at the end of the study, completed questionnaire surveys to provide their subjective opinion of the FIXED 8 schedule. This questionnaire was also administered to non-subjects. In all, more than 100 Submariners completed the questionnaire, and we collected more than 5,000 saliva samples from 24 subjects, and 2000 sleepiness ratings, reaction time, and working memory performance data points from 36 subjects. These are currently being analyzed.
>
> We have also begun analyzing more than 15,000 hours of actigraph-derived data, and obtained specialized software that characterizes sleep episodes using the actigraph data. These analyses will determine the restorative quality of sleep obtained by

crewmembers while following the traditional and an alternative watchstanding schedules, a crucial step toward distinguishing the advantage of the alternative schedule. We also summarized the results of the questionnaires used to glean Submariners' sentiments regarding the two schedules. (162, pp. 15-16).

Subsequently, Loring Crepeau, Chris Steele and Duplessis reported, probably about the SSBN data:

> While the FIXED 8 represented an entirely novel watchstanding schedule, crew members adapted to it with relative facility. They also more positively rated the FIXED 8 over the 6/12 schedule on several factors, including fatigue and energy levels, sleep inertia, and crew morale. (81)

The NSMRL web site displayed the following as I wrote this section in 2014:

> The current 6 hour on / 12 hour off watch schedule adversely affects operational performance and crew quality of life. NSMRL is in the process of evaluating the biologically-grounded, 24 hour-based, FIXED 8 watchstanding schedule (Figure 25 [next page]). This schedule provides watchstanders with the same respective shift every day, in an effort to engender circadian entrainment that leads to enhanced effectiveness during work hours, and more restorative sleep during rest hours. An at-sea evaluation aboard a ballistic missile submarine confirms this schedule's operational employability, and the crew's sentiment toward adopting it was highly positive. Additional data analyses will determine this schedule's influence on circadian entrainment and cognitive effectiveness. A subsequent evaluation aboard a fast attack submarine is planned.

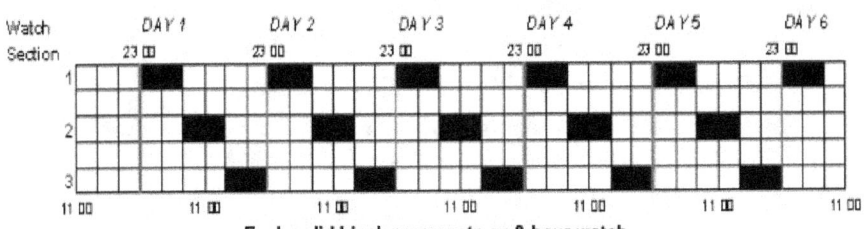

FIXED 8 Watchstanding Schedule

Each solid block represents an 8-hour watch

Figure 25. NSMRL fixed 8 research plan (downloaded 9 December 2014 from http://www.med.navy.mil/sites/nsmrl/Pages/Watchstanding.aspx). See Plans 25 and 35. (I believe that this is a modification of an original figure that I prepared for the laboratory study.)

2008

Michel Paul of the Canadian Forces Defence Research and Development Centre in Toronto (DRDC-Toronto) headed up a limited comparison of three-section *vs.* two-section (6-and-6) watches in a ten-day underway study conducted a Victoria-class long-range hunter-killer submarine (SSK). Gary Gray of DRDC, Tom Nesthus of the FAA's Civil Aeromedical Institute and I helped Michel with the data (247).

Twenty-one subjects participated in the study:

    Three were non-watch-standers

    Six stood a 1 in 2 back-watch: 0100 to 0700 and 1300 to 1900

    Six stood a 1 in 2 front-watch: 0700 to 1300 and 1900 to 0100

    Six stood a 1 in 3 engineering watch

The 1 in 3 engineering watch is shown below as Plan 28 (next page). The actual daily hours worked across the 3-day cycle were 10, 6 and 8, respectively, for a total of 24 hours. There were eight-hour periods off on the nights of days 1-and-2.

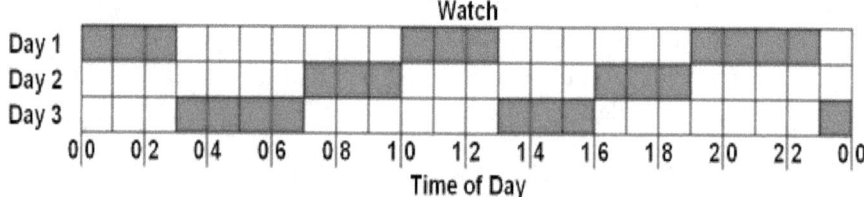

Plan 28. Three-section, counterclockwise-rotating, 4-3-4-3-3-4-3 watch plan (247).

The subjects completed daily activity logs and wore wrist activity monitors (WAM). The latter data were fed into the FAST software to predict fatigue effects. The subjects also filled out pre- and post questionnaires. Michel creates excellent graphs of data, and I have simply copied his

Figures 1 and 3, below. The first figure (Figure 26) shows the prediction for accumulating fatigue across the voyage for the 1 in 2 back watch section. The means for the 0100 to 0700 watch and the minima for both watches were of concern. "The 1 in 2 back-watch-standers had unacceptably low cognitive effectiveness during their 1300 to 1900 hour watch, and potentially dangerously low cognitive effectiveness during their 0100 to 0700 hour watch. Similarly, the 1 in 2 front-watch-standers had low cognitive effectiveness during their 0700 to 1300 hour watch, and potentially dangerously low cognitive effectiveness during their 1900 to 0100 hour watch." The predicted alcohol equivalencies for these estimates were quite alarming. Historically, mariners have performed adequately with these levels of fatigue – most of the time.

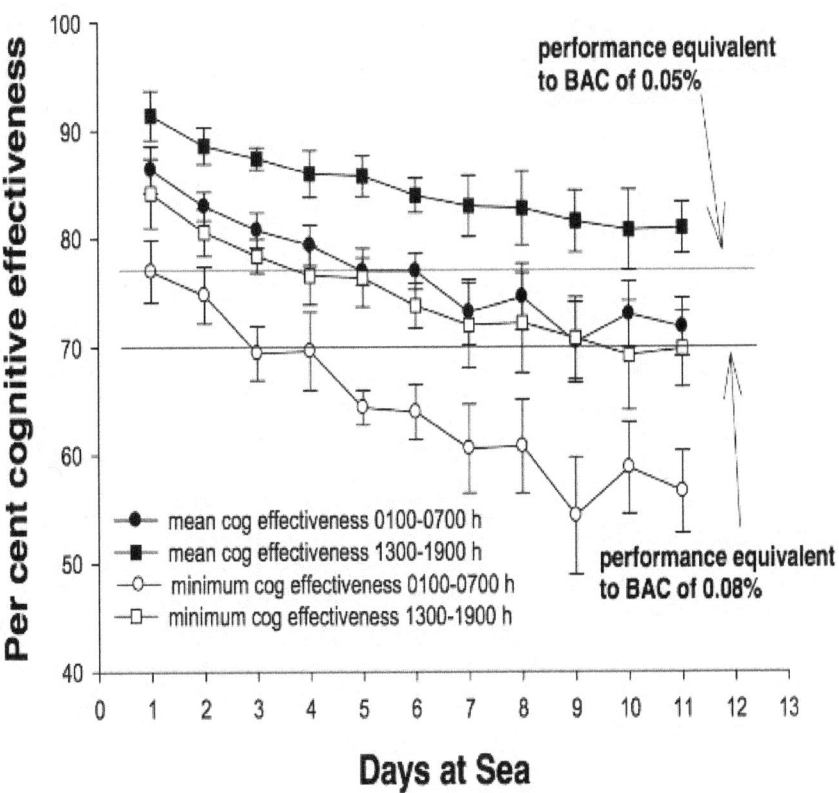

Figure 26. Mean and minimum cognitive effectiveness over days at sea for both of the 1-and-2 back watch periods. Solid circles and squares are mean values ± s.e.m.

and open circles and squares are minimum values ± s.e.m. (247)

My modeling results agreed with the direction of these trends: there was a small day-to-day decline in predicted average cognitive performance effectiveness. My average predicted effectiveness across 15 days of watchstanding (five 3-day rotations) was about 86%, with about 63% of that time spent below my 90% criterion line (BCL). There was a predicted phase delay of about one hour across the first eight days, then circadian stability. One night of ten hours of sleep will be needed for recovery. My modeling probably credited the crew members with more sleep that they acquired in reality, thus my model results would predict better cognitive performance than observed in the real world.

The second figure (Figure 27, below), shows the prediction for cognitive fatigue across the voyage for the 1 in 3 watch section. Due to the watch plan, there were insufficient data to plot predicted cognitive effectiveness across days at sea. However, the mean and minimum cognitive effectiveness for each of the seven watch periods, above, showed that mean cognitive effectiveness was probably greatly impaired for the 1900 to 2300 and 2300 to 0300 watches. In addition to these predictions,

> The activity and sleep log data indicated increasing difficulty arising from sleep and a decrease in subjective levels of 'restedness' over days at sea. Alertness fell over days at sea. Each of the 1 in 2 front and back watches were less happy than their 1 in 3 engineering watch counterparts. While there was no difference in sleepiness between watch system variants or over days at sea, sleepiness levels were consistently elevated to mid-scale levels. Difficulty concentrating, slowed reactions, level of fatigue, work frustration and physical discomfort increased during the trial relative to the pre-trial baseline. (247)

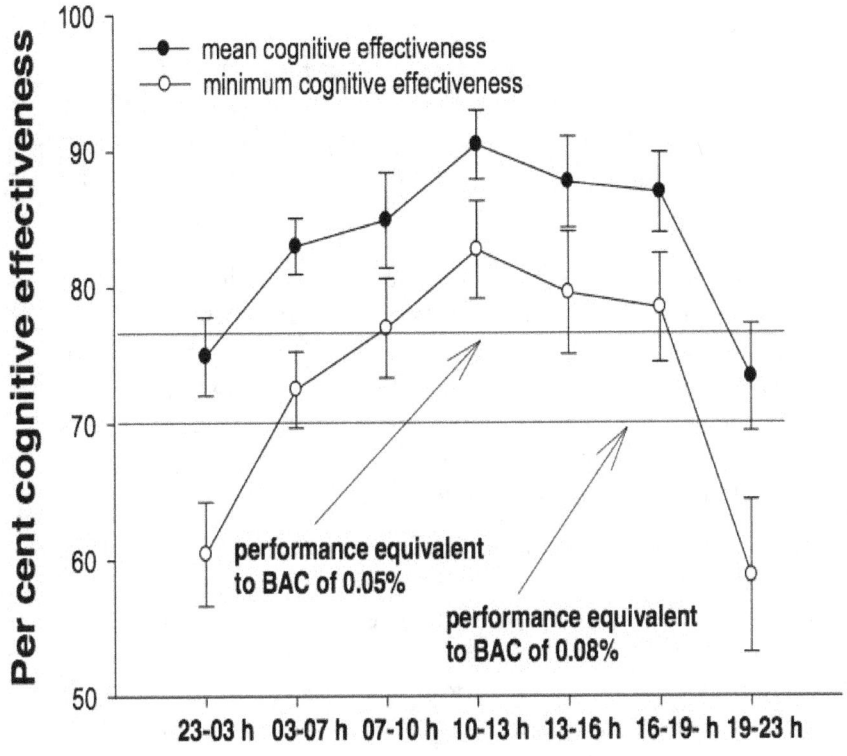

Figure 27. Mean and minimum cognitive effectiveness for each of the seven 1-and-3 watch periods. Solid circles are mean values ± s.e.m. and open circles are minimum values ± s.e.m. (247).

2010

Patrice Baert, now of the École des Applications Militaires de l'Energie Atomique (School of Military Applications of Atomic Energy), Marion Trousselard, now of the Institut de Recherche Biomédicale des Armées (Institute for Military Biomedical Research) and colleagues, posted a summary of an investigation conducted for the French Navy Health Service (CSS/FOST) and the Army Biomedical Research Institute (IRBA). The investigation occurred during and after a 70-day ballistic-missile submarine patrol (11, 304). The watchstanding plan in use during

the patrol is shown below,

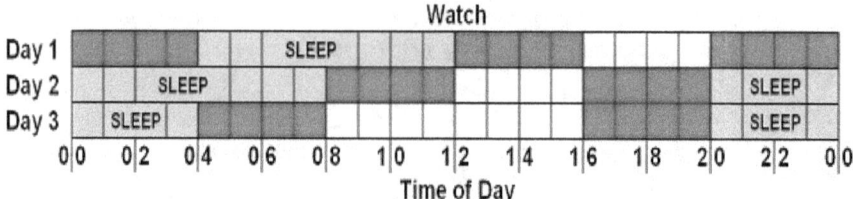

Three-section, counterclockwise watchstanding plan used during the 70-day patrol studied by Baert, Trousselard and colleagues. This plan is identical to Plan 6, above, studied by Colquhoun et al. A dogged version is shown as Plan 18, above, and was studied by Arulandam and Tsing. A modified version was used by the Canadian Navy, shown as plan 28, above, and studied by Paul et al.

The investigators sought data on "cumulative effects of natural light deprivation and of rotating shift-work on psychological and physiological parameters and on cognitive performance." A "second aim was to evaluate the recovery of the physiological, psychological parameters and the cognitive performances two months after the end of the patrol." Twenty submariners provided data "before the patrol (baseline), twice during the patrol, and twice after patrol (one week and two months after)."

An outstanding feature of this investigation was the use of high-quality polysomnography onboard twice while underway, on days 21 and 51. The recordings were made during 8-hour, nocturnal periods. The investigators used a miniature digital recorder (Actiwave, Camtech Ltd, U.K.) and all of the procedures recommended by the American Academy of Sleep Medicine. Polysomnography across the two recording sessions indicated no statistically significant changes in Total Sleep Time (405.6 min and then 392.6 min; about 6.67 hours), sleep efficiency (84% and then 89%) or sleep staging, wake after sleep onset (WASO), etc.

Data indicated that watchstanders were more psychologically disturbed and performed lower on cognitive tests than non-watchstanders. The watchstanders "also exhibited higher cognitive degradations and more sleep disturbances" than non-watchstanders. The physiological, psychological and cognitive disturbances experienced by the non-watchstanders had disappeared one week after the end of the patrol. This was not true for the watchstanders, but their disturbances had disappeared by two months after the end of the patrol.

In their discussion, Trousselard et al. noted, "One possible explanation for the absence of sleep and wake complaints in our subjects despite shift work and the lack of stimulation by natural light may lie in the particular work schedules of these submariners. Established by the [French?] Navy in the 17th century, the schedule allows a minimum of 8 hours of sleep per 24 hours, with a regular phase advance of 4 hours every day and organization of meals and rest synchronized to the shift."

Michel Paul, Steven Hursh and I conducted a small, multi-national symposium on submarine watchstanding plans (249). The symposium was hosted at DRDC-Toronto in September 2009, with attendees from the Canadian Forces (CF), Royal Navy (RN), the United States Navy (USN), the Royal Australian Navy (RAN), and the Royal Netherlands Navy (RNLN). Our objectives at the symposium were "to share submarine watch schedule challenges amongst the participating countries and to model some alternative schedules which would be more sparing of cognitive effectiveness in our submarine crews." We used SAFTE/FAST as our modeling tool.

The U.S. Navy generally uses three-section watch plans in all of its submarines. Additionally, it is likely that ballistic missile submarines use three-section watch plans. According to Trousselard et al., based upon United Nations reports, six countries operate ballistic missile submarines (U.S., Russia, France, U.K., China, and India) and perhaps about 20 to 25 are operational (304). The non-ballistic-missile submarine fleets of the U.K., Canada, Australia, and The Netherlands generally use two watchstanding sections (249). In a two-section plan, each watchstander works 12 hours per day (actual) on every day. This arrangement means that half the crew is manning the ship at any given time. Often, the two sections are called the "Port Watch" and the "Starboard Watch." The two-section solution is quite fatiguing: each section must work at least (168 hrs / 2 =) 84 hours per week. The two-section watches kept on the square-rigged barques or windjammers of the late 19th century and in the British Royal Navy apparently consisted of five 4-hour periods and two 2-hour periods (dog watches). This pattern allowed the two sections to alternate from day to day, so that the Port section had the mid-watch one night and the Starboard section had it the next night (for example, Plan 2).

In the often-used, fixed port-starboard plan, the two sections may alternate at 6-hour intervals or at 12-hour intervals. These plans may start at midnight or at other hours. For example, the 6-and-6 plan may start at 0300, and the 12-and-12 plan may start at 0600. However in some cases, two watchstanders who perform the same task may alternate at less regular intervals. The 6-and-6 plan may be dogged. For example, start at midnight and dog the afternoon watch: 1200 to 1500 and 1500 to 1800. The 12-and-12 plan also may be dogged by breaking the afternoon watch into two 6-hour watches: 1200 to 1800 and 1800 to 0000. Alternatively, one might wish to dog the mid-watch: 0000 to 0600 and 0600 to 1200.

After group discussions, we selected four fixed watchstanding plans for modeling. I describe the four plans next. In these plans, we focused on the idea of when to change the watch during the pre-dawn hours. We looked at two three-section plans and two two-section plans.

Three Section Plans. The first three-section plan used fixed eight-hour watches starting at 0400 (Plan 29). Section 1 stands watch from 1200 to 2000 and sleeps from 0200 to 1000. Section 2 stands watch from 2000 to 0400 and sleeps from 1000 to 1800. Section 3 stands watch from 0400 to 1200 and sleeps from 1800 to 0200. Awakening from sleep occurs two hours before the watch period, providing a sleep recency effect.

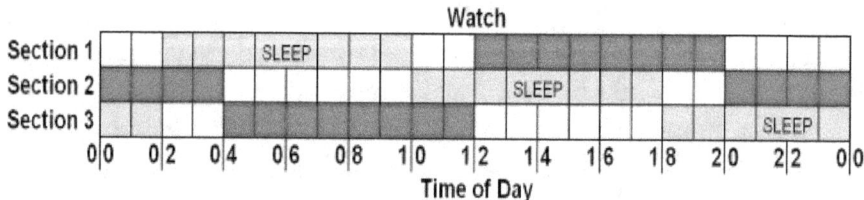

Plan 29. Three-section, fixed, 8-and-16 watch plan starting at 0400 with specified sleep periods (249).

My modeling of section 1 for Plan 29 was not necessary. For section 2, standing the 2000 to 0400 watch, the predicted average cognitive effectiveness was about 84%, with 64% of watch time spent BCL. The circadian rhythm remains stable, and one night is required for recovery. For section 3, standing the 0400 to noon watch, the predicted average cognitive effectiveness was about 92%, with 24% of watch time spent BCL. A phase advance of about two hours occurs over six days, and no nights are required for recovery.

The second three-section plan used fixed four-hour watches starting at midnight (Plan 30). Section 1 stands the 8 to 12 watches and sleeps from 0030 to 0730. Section 2 stands the 4 to 8 watches and sleeps from 2030 to 0330. Section 3 stands the 12 to 4 watches and sleeps from 1630 to 2330. This is the classic maritime watch plan (Plan 1) but with specified, protected sleep periods. I did not model Plan 30 with its protected sleep periods (see Plan 1), but note that six hours of sleep per 24 hours is inadequate for safe operations. Protected napping would be required, also.

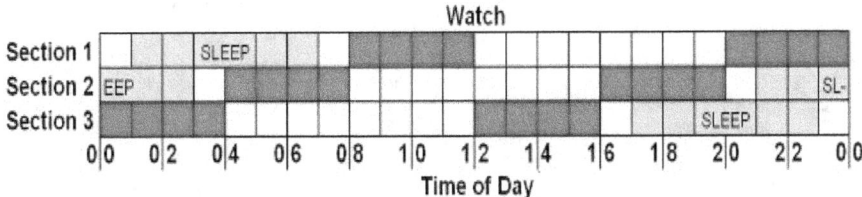

Plan 30. Three-section, fixed, 4-and-8 classic maritime watch plan (Plan 1) with specified sleep periods (249).

Two Section Plans. The first two-section plan considered was a fixed 8-4-4-8 plan starting at 0400 (Plan 31), in which the port section (section 1) stands watch from 0400 to 1200 and from 1600 to 2000, sleeping from 2100 to 0300. The starboard section (section 2) stands watch from 1200 to 1600 and from 2000 to hours, sleeping from 0500 to 1100. This plan was created from discussions among the attendees.

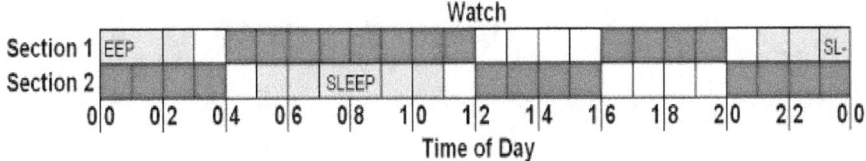

Plan 31. Two-section, fixed, 8-4-4-8 watch plan starting at 0400 with specified sleep periods (249).

My modeling of section 1 in Plan 31 predicted an average cognitive performance effectiveness in watch periods of 90% with 40% of that time spent BCL. A phase advance of about two hours occurs over eight days, but no nights are required for recovery. On the other hand, my modeling of section 2 predicted an average effectiveness of 86% with 66% of watch time spent BCL. A two hour phase delay occurs across 14 days with no stabilization, and two nights of ten hours of sleep are required for recovery.

The second two-section plan considered was a fixed 6-6-6-6 plan starting at 0100 (Plan 32), in which the front section (section 1) stands watch from 0700 to 1300 and from 1900 to 0100. The back section (section 2) stands watch from 0100 to 0700 and from 1300 to 1900. This was the plan in use by CF submarines. The sleep opportunities in this model were based upon sleep data collected during the 2007 at-sea trial on *HMCS Corner Brook* (247). One front watch subjects tended to sleep from 0130 to 0600. One back watch subject tended to sleep only three hours, from about 0830 to 1130. I have modified these observations to allow six hours per day of protected sleep here.

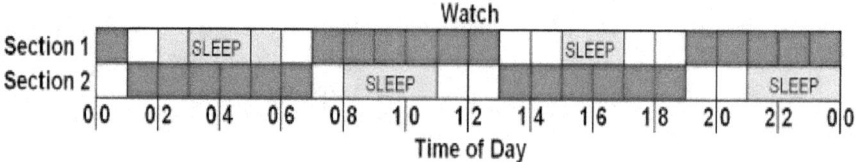

Plan 32. Two-section, fixed, 6-6-6-6 watch plan starting at 0100 with specified sleep periods (249).

My modeling of section 1 in Plan 32 predicted an average cognitive performance effectiveness during watch periods of 91%, with 44% of the time spent BCL. A phase delay of about two hours occurs across 14 ten days, with no stabilization, and three nights of ten hours of sleep are required for recovery. My modeling of section 2 predicted an average cognitive performance effectiveness during watch periods of 81%, with 99% of the time spent BCL. A phase advance of about two hours occurs over about ten days, and one night of ten hours of sleep is required for recovery.

As reported by Paul et al., given the adequate sleep used in modeling, the two three-section plans produced approximately equal predictions for cognitive performance effectiveness during watch periods. Modeling indicated that the two three-section plans, compared to the two two-section plans, "can result in the best levels of crew performance."

Modeling by Paul et al. of the two-section, 8-4-4-8 plan produced predictions of much better cognitive performance effectiveness during watches than did modeling of the 6-6-6-6 plan. This result generated the recommendation, "The proposed new 1 in 2 watch system [8-4-4-8] should be trialed at sea, with a view to adapting operational routines to that watch system." Additionally, Paul and colleagues recommended,

In an effort to further improve sleep and circadian hygiene among submarine crews, appropriate 24/7 lighting (of optimal light energy wavelength and light intensity) could be installed in all appropriate work areas of the boat. During any wake period, when exposure to light is not appropriate, the crew could wear special eye-glasses (e.g., Zircadium(TM) eye-glasses developed by Professor Bob Casper and his team at the Samuel Lunenfeld Research Institute of the University of Toronto) which block out the light energy wavelengths that are responsible for suppression of melatonin. Such a protocol would certainly improve circadian hygiene, which in turn would improve performance. Glasses of this type are now becoming available, and an operational evaluation of their efficacy in long range air transport operations is being planned. The results of that trial will be relevant to submarine operations.

Also,

In those who are unduly tired during a watch, alertness could be facilitated either by caffeinated chewing gum or modafinil, which is an alertness enhancing drug that has been shown in DRDC Toronto studies to be efficacious (255) and have no abuse liability, and which is approved by the US Food and Drug Administration for use by shiftworkers. (249)

Along these lines of thought, Bo Jacobsen, Eva Thoft and Carl Grontmij of Seahealth Denmark indicated that a fixed, three-section 6-and-2 plan might be a good choice (142). The plan is shown below (Plan 33). Each section stands eight hours of watch per day, split into a two-hour period and a six-hour period.

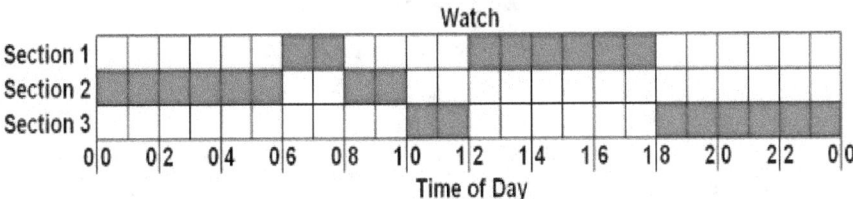

Plan 33. Three-section, fixed, 2-6-6-2-2-6 Seahealth watch plan (142).

As advantages to this plan, Seahealth noted, "All can get long enough sleep. Two get a good night's sleep, someone has to stand watch at night but there is a lengthy period with the opportunity for rest. Shifts match mealtimes." As one disadvantage, I suspect section 2 will have a difficult time sleeping during the day. As another disadvantage, I suspect that crews and captains in a number of different operational environments might object to the frequent watch changes during the 0600 to 1200 period. In support of this observation, see the comment by CAPT Cordle, mentioned above, about "a lot of churn at turnover" (78).

My modeling for section 2 of Plan 33 predicted an average cognitive performance effectiveness during watch periods of about 75%, with all of this time spent BCL. There was a two hour phase delay across all 14 days modeled, with no stabilization, and two ten-hour nights of ten hours of sleep were required for recovery. My modeling for sections 1 and 3 showed no fatigue problems.

2014

In November 2014 the U.S. submarine fleet received a Force Operational Notes (FON) Newsletter, Special Crew Rest Edition. It included good recommendations for sleep hygiene while underway. It used SAFTE/FAST model examples to teach about fatigue effects due to 24/7 watchstanding demands on 6-and-12 and 8-and-16 plans. Appendix B of the FON described a three-section, fixed hybrid 8-and16 plan that looked like this:

Figure 4: Crew Schedule on Hybrid 8s

Figure 5: Supervisor Schedule on Hybrid 8s

Three-section, fixed 8-and-16 plan, Force Operational

Notes (FON) Newsletter, Special Crew Rest Edition, U.S. Submarine Fleet (2014). Green periods are 8-hour watches. Yellow periods are for training, maintenance and qualification. Note all hands awake 0800 to 1600. Blue periods are 8-hour sleep periods.

Appendix B of the FON also described a three-section, fixed, unbalanced watch plan in which one section stands watch during the six hours from midnight to 0600 while the other two sections cover the remaining 18 of the 24 hours (Plan 34). The schedule is "unbalanced" because Section 1 has a six hour per day watchstanding demand while the other two sections have a nine hour per day watchstanding demand. Thus, the plan uses a shift differential based upon hours of work instead of a shift differential based upon hourly pay (196).

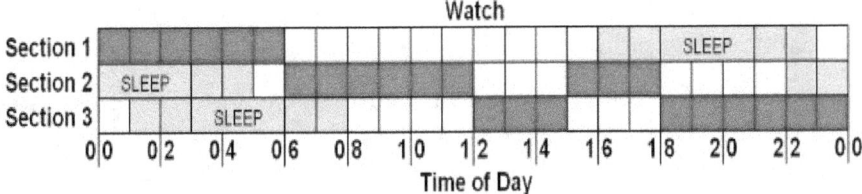

Plan 34. Three-section, fixed, unbalanced, U.S. Submarine Fleet watch plan. Force Operational Notes (FON) Newsletter, Special Crew Rest Edition, U.S. Submarine Fleet (2014).

Not much in the way of modeling is required for this plan. Section 1 will have the usual problems with acclimatization to night work, while section 2, a day shift, and section 3, a swing shift, should have very few problems with sleep or circadian rhythms, given good sleep hygiene.

The suggested seven-hour sleep periods for this plan were Section 1, 1600 to 2300; Section 2, 2200 to 0500; and Section 3, 0100 to 0800. This limiting of sleep to seven hours each 24 hours will induce noticeable cumulative fatigue in a significant number of crew members. I suspect that allowing split sleep would be a better plan for section 1 than a single sleep period to allow a sleep recency effect for the night watch. I have two reasons for this suggestion. First, we know from several well-known experiments and from personal experience that it is difficult for the human brain to generate extended sleep during the daytime hours, even in the evening. Second, I have collected data suggesting that split sleep allows somewhat better cognitive performance in the middle of the night than single sleep periods (210).

2015

Colin Young, Geoffrey Jones, Mariana Figueiro, Shawn Soutière, Matthew Keller, Annely Richardson, Benjamin Lehmann, and Mark Rea of the U.S. Naval Submarine Medical Research Laboratory, the Lighting Research Center of Rensselaer Polytechnic Institute and U.S. Submarine Squadron 6 examined the effects of experimental high correlated color temperature (CCT = 13,500 K; short-wavelength enhanced white light) fluorescent light sources and standard-issue fluorescent light sources (CCT = 4100 K) on watchstanding performance (330). The investigators' hypothesis was that "blue-enhanced" light sources (high CCT) would promote circadian entrainment better than the low-CCT light sources in use aboard U.S. SSN submarines. They noted that similar approaches had been undertaken in Antarctica.

The two types of lighting were used for 11 days each in the engine room aboard the *USS Scranton* (SSN 756) while underway. Watchstanding occurred on a fixed 8-and-16 plan, offset from midnight (Plan 35). Twenty-nine sailors participated in the study, with 10 in section 1, 9 in section 2 and 10 in section 3. Instruments included the Karolinska Sleepiness Scale (KSS), reaction time and matching-to-sample performance tasks, wrist activity monitors, and circulating melatonin estimates. The wrist activity monitors provided estimates of phasor angle, the cyclic phase relationship between an individual's daily rest-activity pattern and the pattern of their daily exposure to light and darkness.

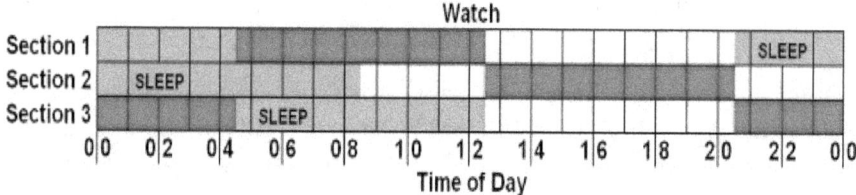

Plan 35. Three-section, fixed, 8-and-16 watch plan from the *USS Scranton* (330). Note that all hands are awake during the 1230 to 2030 period.

My modeling of this plan indicated no real problems for sections 1 and 2. As expected, performance was poor for section 3, working thorugh most of the night. Although average cognitive performance effectiveness during watches was 88%, half the time on watch (49%) was spent BCL. There was a slow, continuous phase delay of four hours that was not complete at 15 days, but recovery was rapid.

The high-CCT light source was associated with significantly greater 24-h behavioral alignment with work schedules using phasor analysis, greater levels of sleep eff ciency measured with wrist actigraphy, lower levels of subjective sleepiness measured with the Karolinska Sleepiness Scale, and higher nighttime melatonin concentrations ... Unlike these diverse outcome measures, performance scores were significantly worse under the high CCT light source than under the low CCT light source, due to practice effects. As hypothesized, with the exception of the performance scores, all of the data converge to suggest that high CCT light sources, combined with 24-h watchschedules, promote better behavioral alignment with work schedules and greater sleep quality on submarines. Since the order and the type of light sources were confounded in this field study, the results should only be considered as consistent with our theoretical understanding of how regular, 24-h light-dark exposures combined with high circadian light exposures can promote greater behavioral alignment with work schedules and with sleep. (Abstract)

SECTION III. FATIGUE MITIGATION
Chapter 11. Modern Maritime FRMS

In my searches of literature relevant to the subject of maritime watchstanding, I have found a number of publications that deal with fatigue risk management in the maritime environment. These publications are considered here under the popular phrase, "Fatigue Risk Management System" (FRMS). Many transportation and other 24/7 industries are implementing FRMS. Usually, an FRMS becomes one portion of an organization's existing or new Safety Management System (SMS), or in the U.S. Department of Defense it becomes a component of Operational Risk Management (ORM).

1974

Don Woodward of the U.S. Office of Naval Research (ONR) and Commander Paul Nelson of the Naval Medical R&D Command published an excellent review of research concerning "the effects of sleep loss, work-rest schedules and recovery on [mental] performance" (328). Their Section II dealt with work-rest schedules (pp. 14-16). Here are some of their conclusions, based upon their review of 57 research publications.

> For periods of up to five days, work-rest schedules of "2 on - 2 off," "4 on - 2 off," "4 on - 4 off," "6 on - 2 off," "6 on - 6 off," "8 on – 4 off," and "8 on — 8 off," can be maintained equally well in terms of performance decrement.

> Work-rest schedules of "4 on - 4 off" and "16 on - 8 off" can be maintained equally well in terms of performance effectiveness for a period of up to two weeks so long as no period of acute sleep loss is experienced and none of the typical "housekeeping" extracurricular tasks have to be performed.

> For periods greater than two weeks and up to 30 days, a "4 on - 4 off" schedule is superior to a "4 on - 2 off" schedule in terms of performance effectiveness. Under normal conditions, performance differences may not be great. Under stressful conditions, however, "4 on - 2 off" and "6 on – 2 off" schedules

tend to result in poorer performance than do those which allow for longer off-duty periods. ...

When a period of acute sleep loss greater than 24 hours is experienced under continuous work-rest schedules of "4 on - 2 off" or l6 on - 8 off," the "4 on - 2 off" worker is most vulnerable, in terms of maintaining baseline performance, and slowest to recover, while the "16 on - 8 off" worker is least vulnerable and quickest to recover.

Regarding the acclimation of the body clock to atypical work-rest schedules, such as watchstanding, they concluded that acclimatization (which they call adaptation) "generally requires a three- to four-week time period."

A period of three to five days typically is required for biological adaptation to night shift work to begin. The process of adaptation is reflected in a flattening of the normal circadian body-temperature rhythm. A complete phase shift, which would indicate rather complete biological adaptation, takes three to four weeks, on the average. There are data which suggests that readaptation to a "normal" day [work]-night [sleep] cycle, after having phase-shifted can occur in three to five days. It would be advisable, in order to take advantage of whatever favorable aspects of biological adaptation there are, to maintain a stable work shift for as long as necessary and not to interrupt it with frequent changes.

Stable day-night shift workers generally perform more effectively than workers who rotate day and night shifts as they would if working a continuous "4 on - 4 off" schedule. There are, of course, individual, cyclical, situational and life-style variables that might equate the conditions, if the variables could be manipulated. We know, for example, that the ability to adapt to night shift work is partly a function of surrounding environmental variables such as the light-dark cycle and the activities of other people.

There is also some evidence which suggests that permanent night shift workers tend to be somewhat different in the cyclic secretion of hormones associated with performance effectiveness. Simply knowing that people differ in their preference for day or night shifts and in their ability to adapt to changes in shift work is important. Even more important is acting on this information by identifying individuals with known preferences, observing performance effectiveness under varied conditions, and using the data gained in making work assignments. For the average worker, night-shift work performance is generally poorer than day-shift work performance; the worst performance occurs between 0200 and 0600 hours. The stabilized night-shift worker, however, performs at night as well as the day-shift worker does during the day; indeed, night-shift workers may perform better if they represent a self-selected group.

Here are some of my thoughts about selecting the right people to be watchstanders. These thoughts were extracted from my *21 Tips* book (205). People differ from one another physiologically in their preferred wake and sleep times in terms of "chronotype." A few may be "larks" (morningness) and a few may be "owls" (eveningness) (133). Most folks (about 95%) tend to fall in the middle of the normal distribution between being an extreme lark or owl. Larks tend to be morning individuals, arising early in the morning and getting to sleep early in the evening, while owls tend to stay up later at night and arise later in the morning. Owls tend to perform better on afternoon and evening shifts. You may assess morningness-eveningness tendencies with the original rating scale (the citation above) or with similar scales available on the Web. From puberty through their early twenties, all individuals apparently have a normal, unavoidable, physiological bias toward being owls (82).

I list here some of the contraindications for starting night work or shiftwork, and the symptoms of maladaptation to this kind of work (280, 296). Contraindications are the conditions and illnesses that suggest that a worker is not well suited for shiftwork or night work. Those that may apply in the selection of maritime watchstanders include:

Conditions

Age over about 45, give or take 5 years.

Extreme "lark;" a person who always awakens by about 5 a.m. and must always go to bed by about 9 p.m.

Rigid sleep time; a person who finds it quite difficult to change his or her normal sleeping hours

Illnesses

> Asthma
>
> Cardiac illness
>
> Depression
>
> Diabetes mellitus
>
> Gastrointestinal illness
>
> Insomnia
>
> Seizures

Maladaptation is marked by chronic and acute conditions and illnesses that suggest that the worker is not acclimating well to the physiological and sociological demands of a shiftwork or night work schedule. They include:

> Acute
>
>> Excessive sleepiness at work
>>
>> Increased numbers of accidents compared to day work
>>
>> Increased numbers of errors compared to day work
>>
>> Insomnia
>>
>> Mood disturbance
>
> Chronic
>
>> Absenteeism
>>
>> Cardiovascular disease
>>
>> Gastrointestinal disease
>>
>> Sleep disorders such as sleep apnea or narcolepsy

1996

Scott Makeig and then-LCDR David Neri, working for the Naval Health Research Center (NHRC) on Point Loma in San Diego drafted a fatigue management proposal (181). Scott and Dave did not cite previous maritime FRMS guidance, and I suspect that they would have been aware if any had existed, at least within the U.S. maritime communities. In their words, they proposed

> an integrated hardware and software system for fatigue and

> alertness management of military shipboard personnel, which would involve: (1) continuous, noninvasive monitoring of crew sleep history via wristband activity monitor; (2) dynamic work/rest scheduling software for optimizing crew schedules under changing missions and personnel demands; (3) real time, objective alertness monitoring of on-duty crew in key work stations using electroencephalographic (EEG) signals recorded via noninvasive dry electrodes built into a cap or audio headset. The system would allow commanders to make operational decisions based on objective knowledge of their crew's state of fatigue and alertness, to maximize human-system efficiency and safety. (181, p. 8)

While the hardware and software approaches suggested here might sound clumsy, there are efficient ways to implement them in military operations (46).

1997

David Neri, David Dinges and Mark Rosekind published a guide under the auspices of the aircrew fatigue research group at NASA Ames Research Center (232). They were responding to questions from the aircraft carrier *USS Nimitz* (CVN 68) and the commander of Naval air operations in the Pacific Ocean (COMNAVAIRPAC) "regarding the human performance effects of a planned 96-hr (4-day)" surge operation.

> The paper is organized into three main sections: (1) background scientific information on sleep and circadian rhythms, (2) responses to operational questions on the effects of sleep loss, and (3) fatigue countermeasures. These sections are followed by appendices that (a) summarize countermeasure recommendations for the surge period, (b) provide fatigue survey guidelines, and (c) list suggested additional sources for more detailed scientific information. (232)

The answers in section 2 of the Neri *et al.* report are of interest here, as the information in sections 1 and 3 is available elsewhere in the present book. Each answer below was expanded upon in the NASA report.

> Q. How long can the crew be expected to go without sleep before significant impairment?
>
> *A. For a crew on a typical schedule with work during the day and sleep at*

night, the critical wake duration after which a majority can be expected to show impairment is about 44 hr (midway through the second night without sleep).

Q. What is the range of individual differences in response to total sleep loss?

A. *There is a wide range. About a quarter to a third of the crew can be expected to show significant impairment after only 20 hr without sleep. Another quarter to third can be expected to show impairment from 20-44 hr without sleep. While the remaining third to half of the crew might be expected to perform reasonably well up to about 44 hr without sleep, all those awake for more than 44 hr will show some significant impairment. Consequently, everyone can be expected to show reduced performance on the third and fourth days of total sleep loss, especially at night and in the morning hours.*

Q. What are the types of problems to expect with extended sleep deprivation?

A. *Sleepiness: Sleepiness will be expressed in a greater probability of unintended sleep onsets in a sleep-conducive environment.*

Physiological and cognitive changes: microsleeps, lapses in [mental] performance, slowed reaction time, reduced vigilance, short-term memory impairment, fixation, poor communication, impaired decision making, loss of situational awareness

Behavioral changes: slowing down (when possible) in an attempt to reduce errors; foregoing routine maintenance in order to perform the primary task

Mood changes: degraded mood, reduced motivation

These performance decrements will increase despite increasing compensatory effort on the part of motivated individuals.

Q. Could the days of reduced sleep leading up to the [surge operation] (SURGEOP) have a significant effect on sleepiness and performance?

A. *Restricted sleep (partial sleep deprivation) in the days prior to the 4-day SURGEOP can have a substantial negative effect on alertness and performance during the SURGEOP.*

Q. Can individuals be relied upon to determine when they are most sleepy?

> A. *A person is not a reliable judge of his or her own level of biological sleepiness.*

1999

In 1999, The Canadian Ministry of Fisheries and Oceans published a 34-page fatigue management pamphlet for the Coast Guard, written by Ron Heslegrave, with Scott Davis and Barbara Cameron of BC Research (131). They cast fatigue and alertness as "two ends of the same scale" and provided bullets about the effects of mental and physical fatigue, and eight symptoms of mental fatigue. The first three causes listed for fatigue were "Working watches," "Overtime or long watches," and "Time of day of the watch, with night watches being most fatiguing." They pointed out the ubiquitous nature of fatigue and gave a one-page tutorial on sleep need (7 to 8 hours) and sleep debt. They also provided a one-page tutorial on circadian rhythms, including the pre-dawn trough and the mid-afternoon (circasemidian) dip in alertness.

The authors included a page on workload, both physical and mental, and they discussed age effects on fatigue susceptibility. They pointed out the problems with working and standing watch at night, noting that there is "no 'best' schedule for watches." They suggested that fixed watch schedules might be best, and that changing the schedule every two weeks was not a good idea. They included tips on stabilizing the circadian rhythm, getting enough sleep, and napping. They suggested behavioral fatigue countermeasures. They discussed the effects of drowsiness-inducing medications, caffeine, alcohol, and diet. They provided tips on managing family life at home. They attached a sleep hygiene checklist and a fatigue and sleep self-assessment form.

This is an excellent pamphlet that has lost very little currency across the ensuing fifteen years. The bullets in the pamphlet could be used as a syllabus for a fatigue management training course for watchstanders.

2001

The year 2001 saw the publication of a number of fatigue management guides. Guides appeared from the U.S. Coast Guard across the years 2001 to 2005, as a result of an effort led by Carlos Comperatore. Skuld and the IMO produced guides, and an Australian Navy Lieutenant offered some excellent fatigue management ideas for submarine operations. All are discussed here.

In February of 2001, Comperatore, Leonard Kingsley and Pik Kwan Rivera of the USCG R&D Center in Groton, Connecticut, published an extensive, bulleted, 110-page fatigue management guide (69). The *Guide* was to be "a resource for the shipping industry to control stress, fatigue, sleep deprivation, and problems resulting from working and living on deep draft vessels." The authors aimed the *Guide* at schedule planners, fatigue management trainers and safety managers. The *Guide* introduced the concept of Crew Endurance Management (CEM). The authors defined "crew endurance" as "the ability to maintain performance within safety limits while enduring job related physiological and psychological challenges."

Comperatore had begun working on a systems approach to crew endurance management with John Caldwell and Lynn Caldwell when they were all at the U.S. Army Aviation Medical Research Laboratory at Fort Rucker, Alabama (68). This was before Comperatore moved to the USCG R&D Center in Connecticut and the Caldwells moved to the Air Force Research Laboratory in Texas (201). For the USCG, Comperatore devised crew endurance management systems for USCG air stations and examined USCG cutter crew fatigue, much as I had (213); I described Comperatore's cutter crew fatigue studies, above (70,75).

In the 2001 *Guide*, the authors described the necessity and formation of a Crew Endurance Working Group (CEWG). They introduced the necessity of and steps for creating the CEWG, labeling it the "cornerstone" of the FRMS: "Controlling risk factors to crew endurance requires the development of a supporting organizational infrastructure. Without management support, individual crewmembers cannot effectively implement endurance management practices." The authors recommended that the CEWG consider at least the following sources of crew fatigue:

> Insufficient daily sleep duration (less than 7-8 hours of uninterrupted sleep)
>
> Poor sleep quality (awakening during the night due to work-related disruptions or noisy environment)
>
> Sleep fragmentation (daily rest periods are numerous but never 7-8 hours of uninterrupted sleep)
>
> Sleep at wrong physiological times (human body naturally

designed to sleep at night)

Changing work/rest schedules (rotations between day and night work once or more times per week)

Long work-hours

No sleep recovery opportunities (napping during the day is not possible)

Poor diet (menu includes frequent fried foods, high fat and sugar content, frequent caffeine consumption)

High workload (high physical and/or mental effort requirements)

High stress (induced by environment, workload, work-schedules, authoritarian leadership style)

Lack of control over work environment or decisions (workers are isolated and not allowed to contribute in problem solving)

Excessive exposure to extreme environmental conditions (cold, heat, high seas)

They described the steps of the CEWG process as:

Step 1. Review crew endurance management information.

Step 2. Identify endurance risk factors.

Step 3. Identify elements affecting endurance during shipping operations.

Step 4. Analyze relationships between elements; determine modifications.

The authors described each CEWG step in detail. I have reproduced below Figure 2 from the 2001 Guide, which the authors described as "an example of results produced by a CEWG during their analysis of a commercial maritime environment. This particular example does not include all elements that may be relevant to maritime environments, but it does include some basic common basic ones. Figure 2 can be used a template for CEWG members beginning the analysis of their particular system." This figure is a compelling visualization of the potential output of a CEWG.

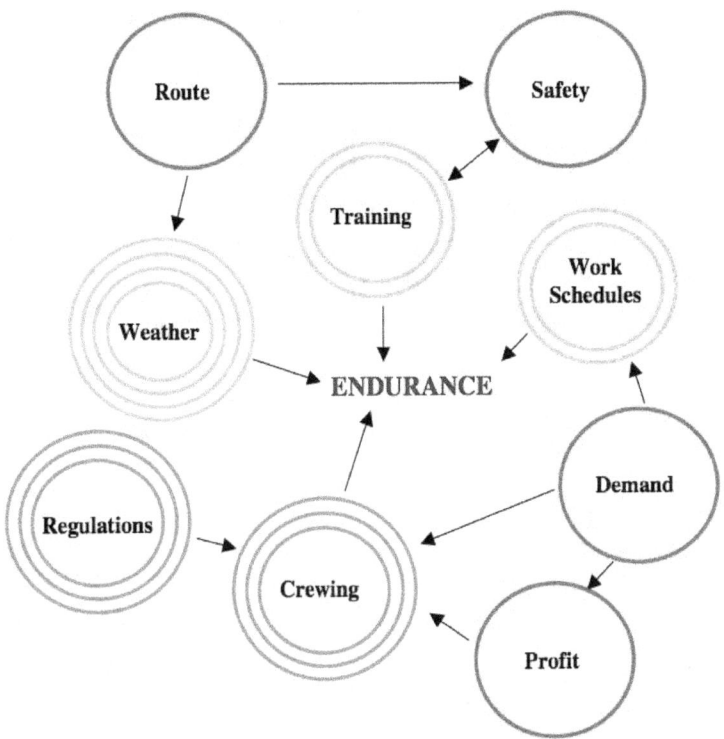

Figure 2. *Example of the system of elements that may impact endurance in shipping operations. A single circle indicates mission objectives, unchangeable elements. A double circle indicates individual actions, elements that an individual or group can modify. A triple circle indicates management support, elements that may be changed through negotiations. A quadruple circle indicates environmental factors. Arrows indicate reciprocal or unidirectional relationships — bi-directional arrows indicate that the elements involved have reciprocal effects.*

Figure 28. Example of results produced by a CEWG during their analysis of a commercial maritime environment (69).

The *Guide* included brief explanations of various fatigue causes such as stress, workload, inadequate sleep, circadian rhythms, etc. The authors included tips for dealing with each specific cause (fatigue countermeasures). The *Guide* also included five appendices: Sleep Management; Napping; Circadian Rhythms; Night Watch Daylight, Sleep, and Work Management; and Shiftwork, Sleep, and Biological Clock Management. The authors updated the *Guide* in 2003, and the newer version should be used; see the reference, below.

233

Later in 2001, the authors published an internal USCG Guide in versions 1.0 and 1.1 (C A Comperatore, Rothblum, Rivera, Kingsley, & Beene, 2001; C A Comperatore, Rothblum, Rivera, Kingsley, Beene, and colleagues, 2001). They updated this document in 2005 as version 2.1, and I have described it, below, under that year.

The Maritime Safety Committee (MSC) of the United Nations International Maritime Organization (IMO) published Circular 1014 on fatigue management in June of 2001 (183). The annex to the circular contained the following nine modules plus an Appendix:

> Fatigue
>
> Fatigue and the Rating
>
> Fatigue and the Ship's Officer
>
> Fatigue and the Master
>
> Fatigue and the Training Institution and Management Personnel in charge of Training
>
> Shipboard Fatigue and the Owner/Operator/Manager Shipboard Fatigue and the Naval Architect
>
> Fatigue and the Maritime Pilot
>
> Fatigue and Tugboat Personnel
>
> Fatigue related documentation

Each module provided technical references for the material in the module. The fatigue factors listed in the first module were of interest because they seemed to be reasonably comprehensive. I've grouped and paraphrased a bit here in an attempt at brevity.

Crew Factors

> Sleep and rest; quality, quantity and duration of sleep; sleep disorders
>
> Biological clock and circadian rhythms; jet lag
>
> Psychological and emotional factors, including stress; fear; monotony and boredom
>
> Health; diet; illness

Skill, knowledge and training as it relates to the job

Personal problems; interpersonal relationships

Ingested chemicals; alcohol; prescription and non-prescription drugs; caffeine

Age

Shiftwork and work schedules; rest breaks

Mental and physical workload

Management Factors

Staffing policies and retention; training and selection of crew

Roles of riders and shore personnel

Paperwork requirements

Schedules -- shift, overtime, rest breaks

Company culture and management style; economics; resources

Rules and regulations

Upkeep of vessel

Voyage and Scheduling Factors

Frequency of port calls; time between ports; nature of duties and workload while in port

Routing; weather and sea condition and traffic density on route

Ship Factors

Ship design; level of automation; level of redundancy

Equipment reliability; inspection and maintenance; age of vessel

Physical comfort in work spaces and accommodation spaces

Location of quarters

Ship motion

> [For those interested in ship design and fatigue, see the thesis by LT Scott Calhoun (47) and the book by Jonathan Ross (269).]

Environmental Factors

    Excessive heat, cold, humidity, vibration, and noise

    Disruption of sleep

    Ship motion--ability to work, nausea, motion sickness.

To this list I add poor personal management of sleep hygiene, i.e., not sleeping when there is an opportunity to do so, and poor design of accommodation spaces to support good sleep hygiene. The latter item refers to the interruption of one or more sleeping crewmembers by others coming off or duty going on duty or being too noisy.

LT Sarah Chapman, a Submarine Psychologist with the Royal Australian Navy serving with *HMAS Stirling*, made comprehensive recommendations for a fatigue management system in submarine operations (49). These recommendations included:

    - Watch systems and scheduling, sleep management and strategic napping in the Collins-class environment;

    - Development, practice and adherence to sleep management plans for personnel;

    - Exploring the inclusion of personnel redundancies for those departments experiencing high cognitive load (to assist in more effective fatigue management); and

    - Modification of task rotation and time of completion to measure cognitive workload and minimize effects of sleep loss (49, p. 4).

2002

In 2002 Lieutenant Colonel P.J. Murphy of the Australian Defence Force Psychology Organisation published *Fatigue Management During Operations: A Commander's Guide*. This excellent review and guide is relevant to maritime operations. Murphy's FRMS guidelines were organized as follows:

    Detecting Mental Performance Deterioration: have folks perform a serial addition or subtraction task for at least a minute, watching for long pauses and errors.

    Improving Wakefulness: Allow frequent breaks; increase social support (teams, buddy system); use periods of mild exercise (if not physically fatigued); identify the purpose of tasks to increase incentive; if possible, change routines and rotate tasks; exposure to sunlight or bright light [properly timed, of course]; introduce novel background noises (such as a radio) for personnel completing mundane or repetitive work.

Enhancing Quality of Sleep: create conditions conducive to quality sleep; minimize alcohol use; manage food consumption; avoid stimulants prior to sleep; if possible, prepare for sleep gradually; ensure all personnel understand the high priority of sleep.

Fatigue Management: What the Commander Can Do: use training to ensure all personnel know their tolerance to sleep loss and to develop the ability to nap; include sleep requirements in operational planning; allow adequate sleep before an operation; monitor sleep periods in yourself, your subordinates and superiors during operations; check performance levels during sustained operations (see test, above); allow each soldier at least four to five hours of unbroken sleep each 24 hour period, preferably at about the same time each day [this is a bare minimum]; when appropriate, adopt a more relaxed leadership style; modify leadership behaviors as fatigue increases, for example, give simpler directions; know the effects of sleep loss; provide environments that facilitate sleep; use strategies to compensate for the effects of sleep loss (see [Murphy] p. 66); promote high levels of physical fitness before operations begin, but do not overtrain once in the field; accept that you have an ethical and operational obligation of self-care. While this is common sense, young and inexperienced leaders often find it difficult to accept.

Fatigue Checklist: see Murphy's Table 8, page 83. I think that his scoring is a bit too liberal. If I were to observe any of these symptoms, I'd want the person to at least take a nap.

Compensating for the Effects of Lost and Disrupted Sleep: use a buddy system where people double check each other [writing, arithmetic, etc.]; permit napping; let members most affected by sleep loss do tasks that are self-paced; adopt a 'brief back' procedure to confirm understanding [a basic communications practice: "What I heard you say was..."]; organize daily workload after a period of sleep to allow for 'mental' tasks to be completed before 'physical' tasks; develop written checklists for operations when levels of fatigue are likely to degrade memory and performance; give priority of sleep to personnel who have critical tasks, whose role or tasks make them more vulnerable to sleep loss, or who are showing that they are more severely affected by fatigue; evaluate the feasibility of establishing separate day and night fighting/operations teams.

Post-combat [or sustained operation] Recuperation: I would rely more on fatigue models than Murphy's recommendations for estimations of needed recovery sleep. However, I agree with Murphy when he reminds the reader that, if exhaustion is severe, unsafe sleeping practices may occur. Supervision may be needed, especially if vehicles or heavy equipment are in the area.

(Doctrine Wing, Land Warfare Development Centre, Department of Defence, Tobruk Barracks, Puckapunyal, Victoria, Australia)

2003

In January of 2003, Comperatore and Rivera of the USCG R&D Center in Groton, Connecticut, updated their 2001 *Guide* for deep draft vessels, aiming this time more generally at commercial vessels (71). I have not reviewed the update here. It appears to be re-organized, updated with a few more references, and improved with materials provided by commercial operators. This updated *Guide* should be used in place of the 2001 *Guide*.

The late Marv McCallum, along with Tom Sanquist, Merrill Mitler, and Gerry Krueger, prepared an excellent reference document for the U.S. Department of Transportation Research and Special Programs Administration (188). The U.S. Department of Transportation's agencies were addressing fatigue management through a multi-modal, coordinated Fatigue Management Program. The *Fatigue Management Reference* created by McCallum and colleagues was a part of that effort. The reference provided basic information that was to be used to "enhance the content of fatigue management guidelines, handbooks, and educational materials." I used the organization of the *Reference* as the basis for the organization of my report for the ASIS CRISP (206). The specific objectives of the DoT reference were to:

- Compile information regarding those factors that can serve as indicators of potential fatigue problems within the commercial transportation industry.

- Describe the basic components shared by fatigue management programs within the commercial transportation industry.

- Provide a series of summaries that address what is and what is not known regarding the efficacy, implementation, and limitations associated with fatigue countermeasures commonly employed in commercial transportation operations. (188)

The authors proposed five FRMS program components:

1. Organizational Commitment. "[T]o succeed, a fatigue management program needs visibility and support at the highest levels in the organization. Organizational commitment requires the allocation of resources sufficient for establishing and sustaining a fatigue management program. Involvement by the upper levels of the organization in supporting, monitoring, and refining the program is required for continued program success."

"The maritime industry was introduced to fatigue management somewhat later than the airline and railroad industries. Following the release of a study documenting the substantial role of fatigue in maritime casualties by the Coast Guard in 1996, there has been a growing involvement in fatigue management within this industry. The Coast Guard leads this effort, providing assistance and guidance through its Crew Endurance program to maritime companies willing to invest in fatigue management." (188, pp. 3-2 to 3-3)

2. Employee-Employer Partnership. "It is a challenge for employees and employers to address fatigue management from a common perspective. At issue are work and rest schedules, which directly affect both operational efficiency and operator well being. On the other hand, because these issues are so critical to the organization and individuals (reducing on-the-job accidents, improving employee health, and improving operational efficiency), they can also serve as an important basis upon which to establish more productive relationships. An effective alertness management program must identify a means of involving both employees and employers in supporting these common objectives. Employees and employers can be most effective in working together jointly in addressing employee fatigue.

"The Washington State Ferry System (WSF) provides a more recent partnership example in the maritime industry. In 2001, at the urging of the

United States Coast Guard, the WSF formed a Crew Endurance Working Group. The group consisted of WSF management, the U.S. Coast Guard (USCG) Research and Development Center team, and employee representatives. The purpose of the group was to study crew endurance and fatigue factors at WSF and to improve conditions that were found. WSF dedicated the funds necessary for the meetings and the ongoing education and training. The USCG R&D team acted as facilitators at the meetings and provided research and data for the group to consider. They have studied various crew work schedules and provided information needed for training and education for fleet personnel. The USCG Marine Safety Office representatives were there on behalf of the USCG, which has regulatory responsibility over working conditions on WSF vessels. Employee representatives supporting fleet personnel have become "Crew Endurance Coaches," and are involved in ongoing training and education." (188, p. 3-4)

3. *Education and Training.* "Education and training provide the knowledge required to support the program at all levels. Education should address the physiological mechanisms that underlie fatigue, provide specific recommendations for countermeasures, and be industry specific. Information should be distributed industry- wide through a range of forums and formats. Information should be provided frequently, in order to foster behavioral change. Updating is required to incorporate new information and techniques.

"Washington State Ferries, in its recent efforts to address operator fatigue, has confronted the basic requirement of educating the workforce prior to introducing major changes in work schedules. One of the first actions of Washington State Ferries' Crew Endurance Working Group (CEWG) was to drastically alter some longstanding work schedules. Although it had been shown that these work schedules had negative health and safety implications, changing them was very unpopular with fleet personnel who had become accustomed to them over a long period of time. The CEWG quickly recognized the need for training and education and had CEWG members visit every crew and provide information regarding proper diet, exercise, and sleep. The WSF Fire and Safety Training book was updated with information regarding Crew Endurance. All new employees and new Deck Officers were trained in endurance and fatigue management during their orientation. Even after this educational effort, there continues to be resistance from rank and file personnel. Change does not come easily to people who feel that their lives have been disrupted; so the training is being expanded and presented to all vessel crews during a one-day seminar on various safety issues, with a segment devoted to endurance and fatigue. The training is being conducted by employee representatives of the CEWG using information provided by the USCG R&D Center team. There has been some very gradual shifting of attitudes from fleet personnel regarding work scheduling and fatigue management, and people are starting to understand the importance of proper sleep, exercise, and diet." (188, p. 3-5)

4. *Employee Health Screening.* "Sleep disorders can disrupt sleep, which can lead directly to operator fatigue, or can contribute to fatigue associated with a challenging work schedule. There are a number of self-screening tools that can be used by operators to determine their own need for further help." (188, p. 3-7)

5. *Program Evaluation and Refinement.* "A fatigue management program requires periodic evaluation and refinement, as in the case of any other aspect of an effective business operation. Program evaluation should be tied back to the established objectives. Measures that could be collected on an ongoing or periodic basis include: hours of charged operator overtime, average number of operator sick days, number of accidents and incidents found to have resulted from operator fatigue, attendance at alertness management educational events, number of operators completing confidential fatigue-screening, and operator responses to a periodic alertness management survey.

"Program refinements close the gap between established objectives and evaluation findings. The process of effectively managing operator alertness inevitably involves coordinating efforts across many members and levels of an organization. Successful refinement requires continued oversight and improvement of the alertness management program." (188, p. 3-8)

2005

Comperatore and colleagues at Groton published an internal USCG Crew Endurance Guide in versions 1.0 and 1.1 (noted above). They updated this document in 2005 as version 2.1 (72). The list of endurance factors used by Comperatore and colleagues in 2005 was much more condensed than the fatigue factors listed a few years earlier by the Maritime Safety Committee (183). The 2005 list was:

Insufficient sleep duration (< 7-8 hrs.)

Poor sleep quality (awakenings)

Breaking sleep into multiple "naps"

Main sleep during the daytime

Rotating between day and night work

Long work days (>12 hr.)

No opportunities to make up lost sleep

Poor diet (high fat, sugar, caffeine)

High workload

High work stress

Lack of control over work environment or decisions

Exposure to extreme environment (cold, heat, high seas)

No opportunity to exercise

High family stress (child and parent care, divorce, finances)

Isolation from family (72, p. vii)

The effects of each factor are explained in the text of the report. Regarding watch plans, the authors discussed:

Traditional Schedules: 1-and-3 and 1-and-6

The 1-and-6 Watch Schedule

Traditional and Alternative Schedules: 1-and-4

Traditional and Alternate Schedules: 1-and-5; and

To Rotate, or Not To Rotate? And How Often?

The authors' traditional 1-and-3 plan, above, is the classic maritime plan that I have labeled here as Plan 1. The 1-and-6 plan is interesting, calling for six watch sections. When I sailed on the USCG's medium and high endurance cutters in the 1990s, frequently the ship had to sail with only two qualified watchstanders for a given position. They would train a third watchstander as quickly as possible. Occasionally, there would be four or five qualified watchstanders for a given position. I do not recall ever hearing of six qualified watchstanders, though it is certainly a possibility.

Coupling this observation with the emphasis on reduced manning in both military and commercial maritime operations, I am hard pressed to see how it was probably practical to implement six-section watch plans on a large scale. However, where this option may be used in critical watch positions, it should be used.

The authors showed that three-section, 1-in-4 and 1-in-5 watch plans tend to disrupt circadian rhythms. They offered an alternative 1-in-4 plan using

> "four watch sections, each section working two 3-hour watches each day. This schedule provides 9 consecutive hours of off-duty time, sufficient for watchstanders to get 7 to 8 consecutive hours of sleep. The hallmark of this schedule is that each watch section stands watch at the same time each day. Thus, personnel can become adapted to a regular schedule, allowing them to be more alert on duty and to get more restorative sleep during sleep periods." (72, p. 5-8)

This plan is different from the four-section, 6-and-18 plans above (Plans 21, 26, 27). It parallels the four-section plan with three-hour watches used by the Roman Army, as described in my chapters on watchstanding history. It is also the basis for the 3-and-9 plan that has received both operational and research support in the U.S. Navy (78, 268, 329). This plan is shown, above, as Plan 20. It seems more practical to implement than a six-section plan. The authors also suggested a five-section method to implement a 1-in-5 plan.

The authors provided the following guidance on watch rotations:

> Avoid using frequently rotating duty schedules.

> If you must use rotating schedules, make sure personnel remain on the same schedule for at least two weeks, and that they rotate clockwise (from 0000-0400 to 0400-0800, for example) rather than counter-clockwise (from 0400- 0800 to 0000-0400, for example).

> Avoid allowing personnel to work more than 12 hours in any 24-hour period. Count the 24 hours from when personnel wake from their normal (longest) sleep period (not naps).

> Allow for self-selection: Give preference to those who need or want to work nights.

The first suggestion runs counter to the successful use of dogged watches for more than 100 years. However, it is consistent with research results that show improved mental performance when the circadian rhythm is stabilized. I disagree with the two-week guidance in the second suggestion. Shiftwork research indicates the superiority of rapid or very slow rotations, such as monthly or bi-monthly over weekly or bi-weekly rotations. There is limited applicability for the self-selection idea in military and para-military operations. However, when applicable, it should be used.

The authors provided detailed guidance on napping and exposure to or protection from bright light for those who work nights. Regarding the use of caffeine, the authors provided the following guidance:

> Consume at least 32 milligrams of caffeine when needed.
>
> Avoid caffeine within four hours of bedtime.
>
> Avoid daily use of caffeine as a stimulant.

My own review of research concerning the use of caffeine as a fatigue countermeasure suggests a slightly different approach, called tactical caffeine use (196, 206). The tactics are:

> Restrict daily caffeine intake to 250 mg or less; even none. To do this, you must be aware of the caffeine content of non-coffee over-the-counter products. Read the labels.
>
> Avoid caffeine within five hours of bedtime.
>
> Use 200 to 600 mg before a critical activity that will occur when you are sleep deprived. The alerting effects will persist for three to six hours. The caffeine gum created by the U.S. Department of Defense (*Stay Alert*) is available commercially as *Military Energy*, and may be practical where liquid forms of caffeine are not.
>
> Note that the diuretic effects of caffeine usually occur with doses of 600 mg and greater.

For additional information, see the Institute of Medicine's monograph on caffeine (67). Keep in mind that all fatigue countermeasures other than sleep are just "band aid" approaches. Recovery from fatigue induced by night work requires sleep. Comperatore and colleagues list some medications that induce drowsiness or alertness.

Berthing is a formidable problem for watchstanders. Single cabins are available to senior personnel on military vessels and for personnel on commercial vessels with minimal crewing. Once two or more watchstanders share the same cabin and are assigned to different watch sections, sleep is likely to be disturbed by roommate comings, goings, entertainment, and social interactions.

Comperatore and colleagues provided suggestions about noise, light, humidity, and bedding in berthing areas. Their most useful suggestions that may be applied to multiple berthing areas were:

> Remove source(s) of noise (for example, disconnect intercom and provide phones or pagers to alert only necessary personnel)
>
> Dampen the source of noise (for example, install door stops or rubber stripping on the door frame to reduce noise from slamming)
>
> Install sound absorbing material (that is, carpeting, heavy curtains, insulated door)
>
> Use ear plugs, but some people experience discomfort that will disrupt sleep
>
> Schedule repair and maintenance activities for late morning or early afternoon
>
> Use heavy dark fabric for curtains
>
> Use heavy dark fabric to drape over door frame
>
> Use eye shades, but some people may experience discomfort that will disrupt sleep
>
>> Note: In some situations, light may be used to adapt crews to night operations and promote increased alertness. Therefore, sources of light must be available in the berthing area. To maximize adaptation and alertness, the intensity of the source(s) of light must be greater than 1000 lux. Recommend equipping berthing areas, and common/recreational areas, with lighting that can attain at least 1000 lux of illumination.
>
> Although there are individual preferences, the ideal temperature for sleeping is 65° F. In multiple person berthing spaces set

temperature at 65° F and provide fans and blankets for members to adjust temperature to their comfort zone.

Keep area clean and uncluttered

Personalize sleep area (that is, photos) to increase comfort

Clean, cool, and comfortable soft linens

Cotton sheets are preferred because they are absorbent, making the practical for any climate, and long lasting

Pillows should be soft enough to conform to the contours of the body

Mattress should support the contours of the body and allow for free movement. Full-size or larger depending on stature of personnel. Allow for at least 6 inches of clearance from feet to end of mattress.

In response to a legislative requirement, The USCG reported progress on the development of their Crew Endurance Management System (CEMS) to the U.S. Congress on 29 March 2006 (308). The USCG conclusion was:

> As shown by the results in this report, CEMS is effective, feasible, and sustainable. The Coast Guard believes that if towing vessel crewmembers and their companies implement CEMS, over time, the crew was probablycome increasingly more alert and will make better decisions. Ultimately, fewer accidents may occur. These same practices and principles apply towards any maritime transportation operation. Accordingly, the Coast Guard recommends that all commercial vessels implement CEMS to reduce the risk of fatigue and endurance-related accidents. (Executive Summary)

2007

Andy Smith of the Centre for Occupational and Health Psychology at Cardiff University reviewed fatigue research and fatigue management with the objective of providing a "basis from which to review the principles for establishing safe manning levels whilst also providing an overview of the broader picture of fatigue in the maritime sector" (293). Smith noted the following, in part (section 6.4):

Safety Management Systems (SMS) are required to include a Fatigue Management System (FMS). A number of the features required in Australian SMS systems are expected to also be effective for managing seafarer fatigue, including the following:

Procedures must be in place to cover the reporting of near misses, accidents, equipment failures, etc. to the appropriate regulatory authority.

A designated person must be responsible for verifying the effectiveness and degree of implementation of the SMS, reporting deficiencies to the appropriate level of management, and identifying people responsible for rectifying deficiencies.

The SMS must be periodically evaluated, and if necessary revised in accordance with documented procedures.

A Check Pilot must be appointed, as part of a continuous improvement process, to observe and make recommendations on individual pilots. The first item on the checklist for Check Pilots is an assessment of the fatigue status of the pilot at the start of each voyage.

Concerning the International Maritime Organization's (IMO) stance on fatigue management (Circular 1014, 2001, above), he wrote,

> The IMO guidelines provide an informative summary of fatigue, yet have a number of limitations which are covered in detail by McNamara et al. (2003) and summarized below (I have removed Smith's examples):
>
> The text of the IMO Guidelines on Fatigue reads more like a general information document than a set of specific guidelines.
>
> Suggestions are often made which may be beyond the control of an individual.
>
> Management are also advised to consider a number of factors thought to influence fatigue, but no specific information with regards implementation is given.

Smith also noted,

> A distinction can clearly be made between personal and operational/legislative fatigue management approaches. Whilst

> both forms of approach to fatigue management have obvious strengths and limitations, the IMO guidelines fall indisputably towards the personal side of this continuum. Given that many seafarers find themselves working in situations over which they have little or no control, such an approach is of little value. It would perhaps be more appropriate to concentrate on operational and cultural change if the issue of fatigue is to be tackled effectively.
>
> Advice and best practice cannot compete with economic pressures. There is often little contingency in terms of crew, as many vessels operate at minimum 'safe manning' levels and are under pressure to complete port turn-arounds quickly. Under such conditions, it appears unrealistic to suggest fatigue-reducing interventions which do not involve some form of economic trade-off, an issue that is not addressed in the IMO guidelines.

To this I add the observation that "operational and cultural change" includes minimizing fatigue effects through the selection and use of optimal watchstanding plans. Economic pressures will cause the selection of plans that call for the fewest numbers of personnel on board a given type of vessel within a given type of operation.

The Norwegian marine insurer, Skuld, published a brief guide, "How to prevent and mitigate fatigue" (290). The guide was "based on the 'Guidance on Fatigue Mitigation and Management' issued by IMO in MSC/Circ. 1014 June 2001." The guide emphasized these points for seafarers and management:

> Ensure compliance with maritime regulations (minimum hours of rest and/or maximum hours of work)
>
> Get between seven to eight hours of deep, uninterrupted sleep per 24 hour day
>
> Take breaks when scheduled breaks are assigned. When possible, take strategic naps
>
> Eat regular, well-balanced meals (including fruits and vegetables, as well as meat and starches), drink sufficient amount of water and exercise regularly.

> Use rested personnel to cover for those travelling long hours to join the ship and who are expected to go on watch as soon as they arrive on board (i.e. allowing sufficient time to overcome fatigue and become familiarized with the ship)
>
> Create an open communication environment by making it clear to crew members that it is important to inform supervisors when fatigue is impairing their performance and that there will be no recriminations for such reports
>
> Schedule drills in a manner that minimizes the disturbance of rest/sleep periods
>
> Assign work by mixing up tasks to break up monotony and combining work that requires high physical or mental demand with low demand tasks (job rotation)
>
> Schedule potentially hazardous tasks for daytime hours
>
> Increase awareness of long-term health benefits from appropriate lifestyle behavior (e.g. exercise, relaxation, nutrition, avoiding smoking and low alcohol consumption)
>
> Both shore and on board management is responsible to ensure that sufficient manning and resources are available so that the requirement to minimum hours of rest and maximum hours of work is met (290)

In about the same year, the London-based International Transportation Worker's Federation (ITF, http://www.itfglobal.org) published a "Fight Fatigue" guide based in part upon the same IMO guidance.

Also in 2007, Nita Miller Shattuck and Captain Robert Firehammer published an award-winning paper subtitled, "The Case for Including Crew Endurance Factors in the Afloat Staffing Policies of the U.S. Navy" (214). The paper summarized the results of research projects by her master's students at the Naval Postgraduate School. I have discussed the results of these projects above. From the abstract:

> The Navy currently uses an afloat staffing policy that is calculated using a 70-hour workweek per sailor metric. However, this construct fails to factor in an individual sailor's capacity to sustain performance and is based instead on a notional Navy standard workweek. Part of the inadequacy of the current staffing policy

results from its failure to consider an inviolable and basic physiological requirement for adequate sleep and rest for sailors. Research indicates a strong causal relationship between sleep and performance. When deprived of sleep, either chronically or acutely, human performance suffers in a dramatic and predictable manner.

From the award citation:

> The paper provides information to correlate workload and sleep conditions with the ability to successfully accomplish operational tasks and the authors recommend watch rotations that consider the ability of sailors to better deliver sustained combat capability. This study and the basic explanations are useful to both policy makers and ship designers in determining the impacts of manning levels for combatant ships, focusing on human performance as a critical component of total system performance. The authors recommend changes in the way the Navy determines manpower requirement for naval combatants, especially considering the trend toward reduced ship manning and the need to perform multiple missions simultaneously.

2009

In his book, *Human Factors for Naval Marine Vehicle Design and Operation*, Jonathan Ross addressed some aspects of fatigue and sleep loss in his chapter 5 (269). He listed as causes of fatigue: boredom, environmental conditions, sleep problems, mental work demands, and physical demands and physiological issues. In the category of sleep problems, Ross listed: loss or disruption of sleep, conflict with circadian rhythms and shift work. I was not especially happy with the latter list, in that the first two problems are caused in large part by the third item: shift work, or watchstanding. While Ross listed many excellent strategies for reducing negative fatigue effects, he made no mention of a careful selection of one or more watchstanding plans for use while underway. In fact, this was just about the only recognized strategy that Ross missed.

Addressing sleep loss, Ross did include as a cause "the standard watch and shift work schedule of four hours on and eight hours off" (p. 63). As strategies for reducing the effects of sleep loss, Ross included good sleep hygiene practices by crew members, the four-section 3-and-9 schedule discussed above, staggered watch rotations so that not all personnel on board reach their circadian nadir at the same time, improved human-machine interface design that takes fatigue effects into account, and the uses of prescription sleep aids and melatonin. While all of Ross' suggestions, except for the prescription sleep aids, are excellent fatigue countermeasures, once again there was no mention of a careful selection of one or more watchstanding plans for use while underway.

While Ross' book was a very fine introduction to human factors for the maritime community, I was disappointed that watchstanding planning received so little attention. Thus, my present book may be viewed as a needed supplement to Ross' book.

Concerning my brief comment above about prescription sleep aids, they are to be avoided for two reasons. First, they are prescribed for use for no more than two weeks. So, they are not a long-term solution for mariners. Second, they induce added cognitive deficits when a crewmember must be awakened earlier than expected, and also for a period after planned awakening (43). I support two alternatives, the first being the use of melatonin. I have studied melatonin in conjunction with studies led by Michel Paul of Canada's Defence R&D Centre (245, 250). It is an effective sleep aid for major sleep periods, as well as being an immune system (174).

2010

Bo Jacobsen, Eva Thoft and Carl Grontmij of Seahealth Denmark, in a publication that I cited earlier, published a fatigue management guide for shipping operations (142). To acquire original information, Seahealth staff "took tours aboard numerous ships and interviewed seamen and shipowners on their experience with watch systems and rest/off-duty time." The authors' topics included fatigue and safety, rest rules (regulation VIII/1 of the STCW Convention), fatigue and health, fatigue and performance at work, watch plans and fatigue, personal countermeasures, cabins and bunks, and diet.

Concerning watch plans, the authors presented an interesting comparison, shown in Figure 29. In general, the comparison supports the idea that there is no "good" 24/7 watch plan: at least one aspect of the watchstander's life will be affected negatively. The exception might be the individual on a fixed day watch.

Watch system	Possibility of getting enough sleep	Length of watches	Distribution of night duties	Link to mealtimes
6-6-6-6	Poor	Medium	Evenly balanced	Easy
12-12	Good	Very long	Great imbalance	Easy
4-4-8-8	Good	Long and short	Equal	Difficult
5-5-7-7	Fairly good	Bit long and short	Option of even balance	Difficult

Figure 29. Comparison of watch plans (142).

Seahealth mentioned a problem to which we seldom attend: interruptions of sleep. They told this little story:

> Engineer Sergey has just finished his evening rounds and is ready to head for bed. He is on-call but just hopes that there are no alarms during the night because he is tired.
>
> It is 02.00 hrs when the first alarm sounds. He gets dressed and goes down to the control room. Luckily, it is nothing serious and he goes back to his bunk 15 minutes later. The same thing happens again at 03.30 hrs.
>
> In the morning he is sitting groggily eating his breakfast. The First Engineer comes and sits at his table: "Morning Sergey, all bright and fit - ready for boiler maintenance?" (142)

Interrupted sleep, when it happens, must be viewed as a major contributor to seafarer fatigue.

Seahealth included a chapter on planning for a change in the watch plan (pp. 22-27). This is the only such guidance that I have found. It was quite excellent. It included a number of factors to consider, including company policies, known and unforeseeable factors (such as weather), crew duties, mealtimes, sleep periods, noise, etc. They provided a detailed approach to coordination of the watch plan with port calls.

Seahealth also indicated that a three-section fixed 6-and-2 plan might be a good choice. That plan was discussed above (Plan 33).

Earlier, concerning the cruise ship watch plan examined by Capt Nick Nash (Plan 19), I mentioned that that he made good recommendations about the interactions between shift planning and the command and control of a ship (226). I quote Nash here regarding that interaction:

> Why introduce something new to a traditional system that has worked very well in the merchant navy for more than 200 years, particularly as I have the luxury of three teams of two officers?
>
> Having just completed a crew resource management and bridge resource management (CRM/BTM) course where fatigue was discussed, particularly for the 12-4, (00-04) watchkeeper, I remembered back to when I was a second mate and the only prospect of getting off the 12-4 was promotion or death – and at that time in the 1980s death looked more likely – so I felt the alternative system might be a way to do this.
>
> Another factor was the introduction of the bridge team command and control (BTCC) concept in P&O/Princess with the functions of navigator and co-navigator driving the ship and command (director of operations) standing back. To move this fully forward it was felt that giving all the officers a chance to take different functions within the BTCC team would be beneficial.
>
> Under the current traditional watchkeeping system the senior team – first officer and a third officer – cover the 04-08; senior second officer and third officer, the 12-04 and the junior second officer and a third the 08-12, traditionally the captain's watch as he is around and about during this period. This makes for an experienced arrival and departure team (4-8 watch) but gives the others little chance to participate in these busy periods. The alternative system will however rotate each team through the arrival and departure watches and so all will get a go.
>
> Another advantage of rotating the officers through each watch on a regular run is that on traditional watches they get used to their 'area of ocean' (for example navigation, currents, traffic, environmental limits and shipboard events such as entertainers using smoke in theatre, passenger bridge visits, crew mess toasters setting off a fire detector at 06:30 and the like.) By rotating them they get to see the bigger picture by operating in all the navigation areas and experience a watch which covers different shipboard events.

This is exactly the kind of thinking needed when one demands nighttime work at a critical task.

2012

One 2012 product of Project Horizon, described above, was a Fatigue Management Toolkit designed for use in FRMS. The toolkit publication (16) was available from Warshash Maritime Academy but required a web search as there seemed to be no link to it. I found it at http://www.warsashacademy.co.uk/about/resources/horizon-fatigue-management-toolkit.pdf.

The toolkit was composed of two parts: recommendations and a fatigue modeling tool called MARTHA. The recommendations, each of which included action items for various actors in fatigue management were:

> Use a Sleep Wake Predictor, for example MARTHA, within a fatigue risk management system.
>
> The off duty period on the 6-and-6 watch plan must be reserved for sleep and rest.
>
> Consider two persons on watch on the bridge at the same time. (This gives both physical support (keep each other awake) and mental (two tired persons make better decisions than one person alone).
>
> Education on all levels.
>
> Understand the effects of new shift patterns [plans].
>
> Monitor the 6x6 watch system carefully. (For example, it may be better to change watch at 03-09-15-21, than at 00-06-12-18. If it is necessary to rotate the watch, it should be moved forward.
>
> Consider the design of the sleeping environment.
>
> Include Fatigue Risk Management Systems (FRMS) within the ISM Code [International Safety Management Code]
>
> Consider developing regulatory systems for the management of fatigue, to create a chain of responsibility in the transport sector.
>
> Create individual solutions when assigning watch.
>
> Promote healthier lifestyles for seafarers: food, exercise, stimuli, light, socially (and at home)

> Educate seafarers and managers on the importance of strategic sleep and sleep hygiene.
>
> Consider "Caffeine naps" of 10-15 min. Drink a large cup of coffee and lay down for 15 min. The caffeine and the nap have an interacting effect.
>
> Consider Naps of < 45 or > 90 minutes.

I agree with all but the last of these recommendations. Instead of the last, I teach that "any sleep is good." The premise for nap length limitation was discarded quite some time ago as we are unable to predict what stages of sleep will occur, nor when, within a nap. Sleep staging within naps depends upon cumulative sleep debt, stage-specific sleep debt, length of prior wakefulness, time of day, and other factors, none of which can be monitored reliably during operations. Thus, there is no way to predict an individual will enter deep (slow-wave or REM) sleep after 45 minutes, nor depart deep sleep after 90 minutes. Nor can one predict that a classic 90-minute sleep cycle will occur during the daytime.

The second part of the toolkit describes a publicly-available "device for prediction of sleep/fatigue under maritime conditions (MARTHA). The 'MARTHA' acronym is derived from 'a maritime alertness' regulation tool based on hours of work.' "

> The development of Martha used the basic modeling concepts described in the previous section. The functions were combined and provided with a maritime interface with selectable watch schedules, a do-it-yourself watch system facility, a six week time window for prediction, estimates of time at risk and time with fatigue for each watch and for the whole watch schedule, as well as for time outside watch duty. The major display contains estimates for each 24h period, but there is also a second display to describe each 24h period – with sleep periods and a continuous estimate of sleepiness. This information may also be displayed as miniatures in the main display. (p. 17)

In the Appendix, I describe my use of the SAFTE/FAST fatigue prediction model to assess the relative qualities of the watchstanding plans presented throughout this book. The underlying models used to build MARTHA and SAFTE were quite similar. The following *Toolkit* text describes the underlying model for MARTHA.

Mathematical models for alertness or performance prediction have been developed mainly as tools for evaluating the effects on sleepiness or fatigue of work schedules or sleep/wake patterns that deviate from the pattern of day time activity and night time sleep. The first (two-process) model, however, mainly addressed the question of sleep regulation and by inference, the need for sleep or fatigue/sleepiness (Borbély, 1982). Its main components were one factor representing the homeostatic effects of time awake and amount of prior sleep as well as a circadian component representing the effect of the biological clock. The homeostatic factors are usually seen as having an exponential relation to sleepiness. This means, for example, a steep initial fall of alertness (or rise of sleepiness) after awakening, with a gradual flattening out towards an asymptote of very low alertness after 24h of time awake. The effect of sleep shows a pattern of restitution that increases post sleep alertness. The circadian component is usually represented as a sinusoid function with a 24-hour period which results in high.

The first model to focus explicitly on sleepiness was inspired by the Borbély et al model but also included sleep inertia. It was called The Three Process Model of Alertness (TPM) (Folkard and Åkerstedt, 1991) (Åkerstedt and Folkard, 1995). ...

Process C represents sleepiness due to circadian influences and has a sinusoidal form with an afternoon peak. **Process S** is an exponential function of the time since awakening, is high on awakening, falls rapidly initially and gradually approaches a lower asymptote. At sleep onset process S is reversed and called **S'** and recovery occurs in an exponential fashion that initially increases very rapidly but subsequently levels off towards an upper asymptote. Total recovery is usually accomplished in 8 hours. As discussed below, this recovery function is too steep and permits too rapid recovery when sleepiness is increased because of sleep loss. The exponential function increases in steepness of slope with increasing sleep loss (i.e. low S). **Process U** represents a low-amplitude rhythm with a 12h period, with its acrophase locked to 3h before the circadian acrophase. The use of this component is not validated but is retained in most development and validation work described below. The final component (not in fig) is the wakeup **Process W**, or sleep inertia. The latter is not relevant for the present purpose since sleep inertia is only present for 15-30 minutes after awakening. (pp. 14-15)

The three-process model is shown in Figure 30. The main difference between the underlying models for MARTHA and SAFTE was that SAFTE included Process W, the sleep inertia function. The reason for the latter is that SAFTE was designed for military operations, in which there is concern about alertness upon sudden awakening.

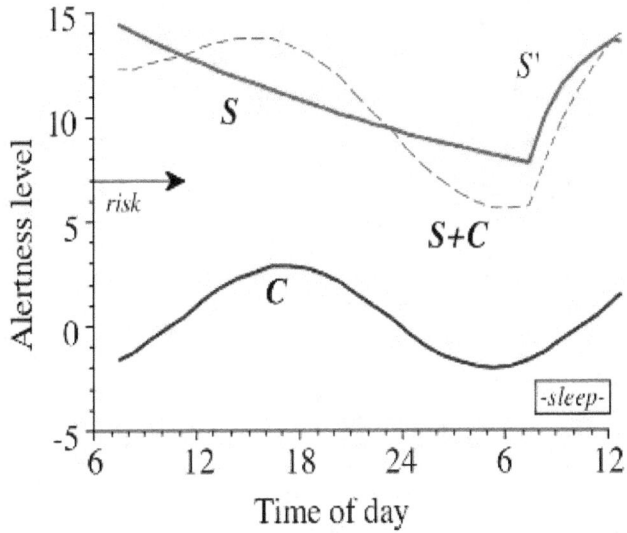

Figure 1. The components of the three-factor model of alertness. See text for explanation.

Figure 30. Three-factor alertness model from the Horizon *Toolkit* report (16, p.16).

I liked MARTHA. It was a good idea and had a pretty nice interface (Figure 31). The supporting literature needed more emphasis on the work that had correlated the Karolinska Sleepiness Scale (KSS) with EEG. Not all of the watch length options provided by MARTHA would fill its entire six-week calendar with a single watch plan by using a single key-stroke combination. Only the common plans used on commercial vessels would fill the calendar. These were the 6-and-6, 4-and-8 and 12-and-12. The other watch length options required entry on each day in the six-week calendar. Once a plan was entered, there was no option to save it. Also, once I had entered a plan, when I came back to day 1 in the calendar I saw that the sleep period on day 1 had been modified. I had no problem with the modification, itself, which I assumed was model-based. However, as a user I could not see easily that it had happened nor why it had happened. Mike Barnett of Warshash informed me that the Project Horizon folks hoped that a commercial enterprise would take over MARTHA and develop it further. I hope that happens.

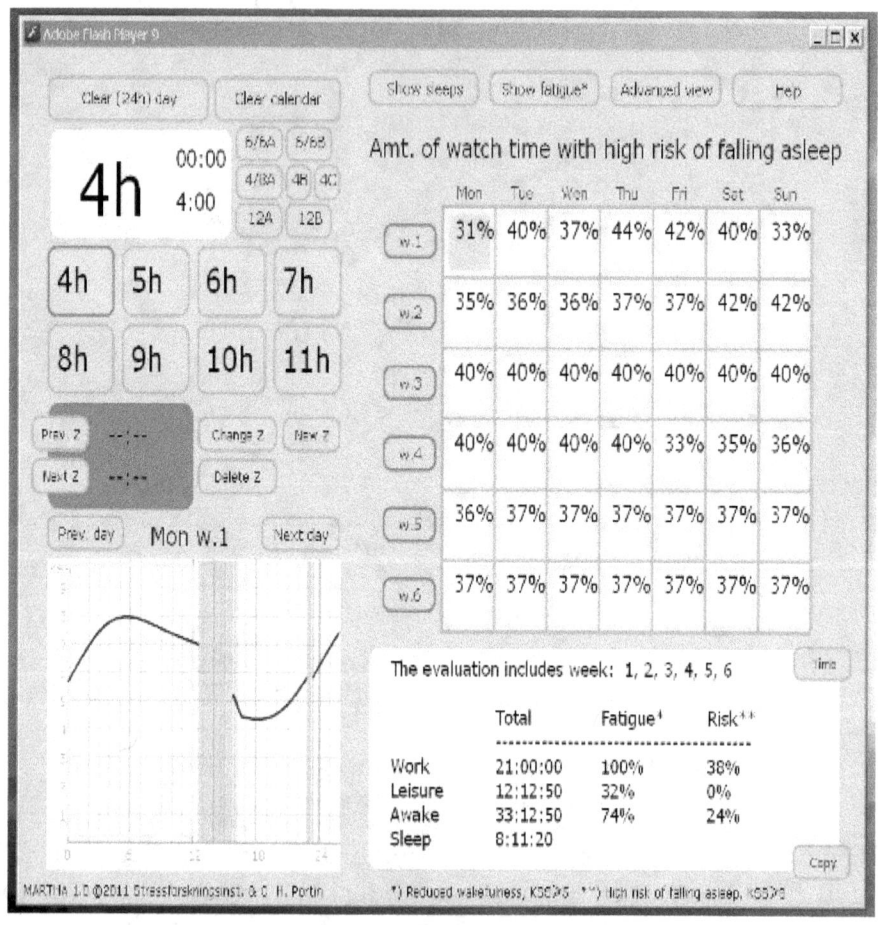

Figure 31. Screen shot of MARTHA interface.

In his U.S. Naval Postgraduate School masters thesis, LT Donald Roberts created a fatigue management guide for commanders (268). This guidance was based upon his study of the data from the *USS San Jacinto*, described above. Roberts recommended the following watch plans:

Having only enough qualified watchstanders for two sections, use a fixed 12-and-12 plan.

With three sections and no shifting, use the classic maritime plan [Plan 1].

With three sections with shifting, use a 6 and 12 plan [Plan 23].

With four sections and no shifting, use a 6-and-18 plan [Plans 26, 27].

With four sections with shifting, use a 3-and-9 plan [Plan 20].

Roberts also wrote a "Commanding Officers' Watch Reference Guide" (268, Appendix B). In his Guide, he compared several four-section plans in terms of SAFTE/FAST-predicted cognitive effectiveness, and did the same for two-section plans. The comparisons supported his recommendations, above.

2013

Australian Marine Order 54 had required that all maritime pilots operate within a fatigue risk management plan (FRMP). The Australian Maritime Safety Authority (AMSA) published a default plan per provision 58 of Marine Order 54 (9). They scored two different kinds of passages performed by a pilot as having either 1 or 2.5 points. Interestingly, they specified a maximum 28-day roster cycle. Twenty-eight days is a cycle length used frequently in shiftwork plans; it divides the shiftwork year of 364 days into thirteen equal periods (196).

AMSA specified required rest periods for pilots that were based upon passage type, upon points accrued per unit time and upon sequential numbers of days worked. Rest periods were, essentially, nights spent at home. A roster reset was generated upon five consecutive nights spent at home. The numbers used in this default plan were instructive for marine pilot operations and also for shiftwork planning.

2014

Ross Owen Phillips of the Institute of Transport Economics in Oslo, Norway, called for a pan-modal (my term) view of fatigue throughout transportation research (254). I usually get an ironic laugh during lectures when I address the question, "Why would fatigue research in that industry be applicable to my industry?" I like to answer, "Well, it's only applicable if your operators are human." I have tried here to paraphrase the points made by Phillips in his Conclusions section.

> Our understanding of fatigued states in transport operators is limited by studies using unique customized measures.
>
> Understanding would be improved if applied studies were to use a standard measure battery concerning perceptions of sleepiness and fatigue. He suggested:
>
>> Epworth Sleepiness Scale [I use this once per study as a screening instrument]
>>
>> Swedish Occupational Fatigue Index
>>
>> Karolinska Sleepiness Scale
>>
>> Samn-Perelli scale [Actually, the USAFSAM Fatigue Scale (201)]
>
> Tables of average scores and shares of samples scoring above threshold scores on each constituent battery measure.
>
> Standardization of physiological measures, plus a focus on naturalistic observations of operators in real world situations.
>
> More research on the interrelated health and safety effects of accumulative fatigue.
>
> Be explicit about the operational definition of fatigue, especially in relation to sleepiness.

Phillips noted the following about the assessment of the fatigue process:

> Working time and sleep quantity and quality have been relatively well studied in all sectors. Road studies are more concerned with charting work and sleep patterns in relation to general fatigue levels, while rail studies tend to focus on the acute effects of schedules on fatigue at work. Sea studies are characterized by a focus on the effect of watch systems on sleep and fatigue.
>
> There is increasing recognition that the quality of work may be just as important as the quantity of work (i.e. work time) in terms of resulting fatigue.
>
> Comparative studies of the different conditions faced by operators in different subsectors ... would help illustrate the fatigue-related challenges to regulators and organisations.

The quality of life outside work has been overlooked as contributor to fatigue at work.

Phillips noted the need for,

> Improved reporting of near misses and accidents in all sectors. [For assistance, see my ebook, *Anatomy of a Fatigue-Related Accident* (197).]
>
> Standardization the periods for which operators are asked to report on incidents of severe sleep (or falling asleep) while operating. [The two-hour epoch works well in many settings.]
>
> Increased use of operational parameters, such as brake and accelerator use, could help chart the effects of fatigue on real performance. [Unfortunately, operational systems have few, if any, electromechanical "hooks" form which an investigator may acquire data.]
>
> The way in which fatigue influences more complex aspects of performance (e.g., increased reliance on default mental schemas). This requires that such performance effects can be operationalized for study.
>
> Consideration of the costs of fatigue to the long term health of the operator and to performance of safety tasks that are secondary to the main operator task.

Finally, Philips noted,

> [W]e have expanded Dawson and Fletcher's (2001) five-level fatigue-risk trajectory to account for fatigue determinants beyond working time, and recovery from fatigue beyond sleep. ... A consideration of the findings of the studies retrieved for this report in relation to the modified trajectory reveals the following:
>
>> - While work time has been well studied, there is a need to consider how work time, work quality and non-work life quality interact to influence operator fatigue.
>>
>> - It has been established for many operators that restricted sleep and perceived fatigue results from work time demands, and that restricted sleep impedes recovery. However, there is much less consideration of recovery during non-work wake time, which evidence suggests is

important. Such recovery could be assessed using a measure of need for recovery following work and preceding subsequent work periods (291).

- The fatigue assessment battery described above, supplemented by the identification of fatigued operator symptoms and behaviours, could help trigger countermeasures that prevent operator fatigue affecting performance.

- Considerations of main fatigue predictors based on sleep history, time on task and time of day are the basis of software parameters allowing schedulers to predict fatigue risks. However there is little understanding of the dynamic interactions between poor sleep and health and psychosocial pressures, which may lead to risks that are poorly predicted by existing software.

- Longitudinal studies on the effects of accumulative fatigue and burnout on operator performance and attrition from the industry are required.

2015

Anna Anund of the Swedish National Road and Transport Research Institute (VTI), and colleagues, produced a countermeasures report that covered fatigue issues in road, rail, sea, and aviation (6). They wrote an interesting introduction to the "Sea" section of the report:

> The ship as a working place is exceptional and by no means comparable to the working places in other modes of transportation, as was already put forward in the 1950s (Aubert & Arner, 1958):
>
>> The ship is a total institution where the seafarer lives at his place of work, among his colleagues and superiors;
>>
>> The seafarer is physically isolated from the family for considerable amounts of time.
>
> Such unique circumstances will undoubtedly influence fatigue and its possibilities of mitigating it as it does increase the psychological stress in seafarers (Carotenuto, Molino, Fasanaro, & Amenta, 2012). On the one hand, seafarers do not have domestic duties in

the same way as those who live at home. On the other hand, worry over family matters at home might also be a cause of stress for seafarers. (p. 27)

They summarized the work-rest requirements of European council directive 1999/63/EC, applicable to "seafarers on board every sea vessel registered in the territory of a Member State":

> either the maximum hours of work which must not exceed:
>
>> 14 hours in any 24h period
>>
>> 72 hours in any 7-day period [an average of 10.25 hours work per day]
>
> or the minimum hours of rest which must not be less than:
>
>> 10 hours in any 24h period
>>
>> 77 hours in any 7-day period [an average of 11 hours off per day]

They also summarized the maritime labor convention of 20 August 2013, ratified by 64 states as of August 2014. "The convention applies to all ships, irrespective of size, except fishing boats, handmade boats and warships." They also noted the IMO STCW Convention (Standards of Training, Certification and Watchkeeping of Seafarers), which is very similar to the MLC requirements.

They described one technical solution, the Bridge Navigational Watch Alarm System (BNWAS). The system "has a dormant stage together with 3 alarm stages" and it is activated automatically upon activation of the autopilot.

> Stage 1: Upon engagement, the bridge officer is required to signal his presence to the BNWAS every 3 to 12 minutes in response to a flashing light, either by moving an arm in front of a motion sensor, pressing a confirmation button, or directly applying pressure to the BNWAS centre.
>
> Stage 2: When a confirmation signal does not occur within 15 seconds, an alarm will sound on the bridge; if there is still no confirmation signal after an additional 15 seconds, an alarm will sound in the captain's and the first officer's cabins. One of them must then go to the bridge and cancel the alarm.

> Stage 3: If neither the captain nor the first officer cancels the alarm within a specified time period (between 90 seconds and 3 minutes, depending on the size of the vessel), an alarm will sound in locations where other personnel are usually available. (p. 29)

They noted that the International Maritime Organization (IMO; http://www.imo.org) had published guidelines on fatigue and its mitigation and management. They a;lso noted that the Nautical Institute had published a few fatigue management tools on their website at http://www.nautinst.org/en/forums/fatigue/fatigue-management-tools.cfm.

> Those included:
>
>> ISF Watchkeeper software. This software is designed to show whether the working hours of crew are in line with the hours of rest regulations;
>>
>> Crew Endurance Management System (CEMS). This is a tool developed by the US Coast Guard and enables companies and crewmembers to manage the occurrence and effects of crew endurance risk factors (such as fatigue) that can lead to human error and performance degradation.
>>
>> MARTHA – a new horizon. MARTHA is a prototype fatigue prediction software model, available through the website of Warsash Maritime Academy (UK). Its purpose is to optimise operation and work schedules by minimising average fatigue predictions (http://www.ship-technology.com/features/feature-project-martha-reducing-seafarer-fatigue/)
>>
>> Maritime New Zealand – Fatigue management. Maritime New Zealand has collected a wide range of resources to help seafarers and managers in the maritime industry to better understand and manage fatigue. Among links to different sleep(iness) tests and questionnaires is also the booklet "understanding fatigue – get your sleep, reduce your risk". (pp. 30-31)

They also summarized briefly some research about causes of seafarer fatigue including traveling to the ship, port visits, ship design, working night time rather than daytime, abrupt changes in work schedule, watch systems, duration of tour of duty, i.e. time at sea, and work-related factors.

On the subject of the prevention and management of fatigue, they suggested the uses of reactive and proactive measures (Starren, van Hooff et al. 2008):

> The reactive countermeasures that were found by Starren and colleagues are in line with those described in the IMO guidelines mentioned previously, whereby napping and strategic caffeine consumption where by far most frequently reported by a group of international maritime experts.
>
> Proactive countermeasures serve to prevent the onset of fatigue and the ones that Starren and colleagues have found here mostly relate to sleep and sleep hygiene, whereby the top-5 consists of 1) a good sleep environment, 2) 7 to 8 hours uninterrupted sleep per night, 3) 2 consecutive nights recovery sleep, 4) adequate sleep, quality of sleep, and 5) obtaining the same amount of continuous sleep as normally at home (Starren, van Hooff et al. 2008) (p. 32)

One concluding note was that "Although the most frequently reported countermeasures (e.g., napping and coffee) have proven to be successful in other contexts, they have not been systematically investigated in the maritime context."

Chapter 12. Concluding Thoughts on Watchstanding Research
[ToC](#)

Generally, research on watchstanding plans has provided pretty much the same picture as research on various 24/7 shiftwork plans. First, working between midnight and dawn generates acute fatigue and concomitantly elevated relative risks for errors, incidents and accidents. Conversely, sleep acquired during the daylight hours will be less restorative than sleep acquired during the nighttime. This will often lead to cumulative fatigue and concomitantly elevated relative risks for errors, incidents and accidents. These are inescapable biological facts that lead to the need to implement fatigue risk management in 24/7 operations.

The cumulative fatigue issue can be "interesting." My use of the two-process model to examine shiftwork plans *a priori* and *post hoc*, and accidents *post hoc*, has shown me some counterintuitive aspects of recovery from night work. It is now obvious to the casual observer that there will be fatigue issues in work that occur during the pre-dawn hours. It is less obvious that the unexpectedly slow recovery from cumulative fatigue will lead to workers being fatigued at the beginning of daytime work periods following days off that have been allotted for recovery from night work.

In the domain of shiftwork, this has led to varying guidance. I helped the U.S. Pipeline and Hazardous Materials Safety Administration (PHMSA) write the following guidance for controllers of gas and oil pipelines in the U.S. This guidance was produced in 2011-2102 first by use of the two-process model, then modification through in-depth discussions undertaken by the PHMSA Control Room Management (CRM) Team, of which I was a part, and the operators of six different control rooms in the U.S. pipeline industry. Thus, the guidelines incorporate both theory and practice. They advise that reasonable maximum normal limits on controller hours of service should be:

> Sixty-five (65) duty hours in any sliding 7-day period. Usually this would be scheduled as seven 8-hour shifts, six 10-hour shifts or five 12-hour shifts.

> Fourteen (14) duty hours in any 24-hour period, which includes shift hand-over time.

At least thirty-five (35) continuous hours spent off-duty when any one or more of the following limits is reached:

Shift starts on seven successive days or nights;

65 duty hours in any sliding 7-day period;

Seven 8-hour shifts in any sliding 7-day period;

Six 10-hour shifts in any sliding 7-day period; or

Five 12-hour shifts in any sliding 7-day period.

35-hours off may be used as a "reset" within any sliding 7 day period if and only if it follows a sequence of two or more day shifts. For example, the 12-hour DDDONNN sequence is acceptable even though it appears to violate the 65-hour HOS guideline (6 days x 12 HOS per day = 72 HOS in 7 days). The day off in this sequence begins in the evening and extends 48 hours to the beginning of the next night shift, providing the opportunity for two nights of sleep.

How might these guidelines apply to watchstanding while underway? First, the 65-hour limit in seven days allows three-section plans with eight hours of watch per day (56 hours of watch per week) plus nine other hours of work.

Although the limits of 14 hours per day of duty time does allow crewmembers to stand twelve hours of watch per day, the 65-hour limit does not allow two-section plans with twelve hours of watch per day (84 hours of watch per week).

Often, dogged watches and four-section plans are useful because they induce watch rotation. The rotation helps stabilize the circadian rhythm when watchstanders' body clocks are affected by a non-changing daylight/darkness *Zeitgeber*.

For below-decks workers, the three-section, 8-and16 plan may work well. This is where the use of light therapy and light protection might be quite useful.

Flexibility is good, especially in a two-section watch plan. For example, in a two-section 7-5-5-7 plan with watch turnovers centered on meals at 0730, 1230, 1730 and 0030, one commander reported allowing the forenoon watch keepers to eat lunch and handle their cleaning responsibilities in shifts prior to the end of the watch. This allowed them to get a meaningful rest period in the afternoon.

There is no good two-section watch plan. In fact, there is no good plan for 24/7 ops on land or on sea, with one possible exception. Because of sleep drive/sleep homeostasis and circadian rhythm effects, whomever must work instead of sleep from midnight to dawn will suffer short-term impairment and, in some cases, long-term health impairment. Thus, some sort of rotating watch plan may be advisable. The exception that I have in mind is for below-decks crew and submarine crews. If they use present research knowledge and biotechnologies to entrain their body clock to their watch schedule carefully, they may work a fixed watch plan without much impairment.

Port visits can be important for recovery from watchstanding. However, the watchstanders would have to be relieved of in-port duties to allow them to sleep.

Non-watchstanding responsibilities for military personnel continue to be a problem with respect to obtaining adequate sleep while underway. This is an age-old saga. Notice that many commercial tankers and container ships have reduced their crews to very small sizes. They use a lot of automation, have no administrative or training requirements underway, as do military vessels, and also do very little ship's maintenance while underway. This may be the only model available to us for fatigue mitigation on military vessels. Perhaps there is a way to organize the periodic port call such that sleep, training, admin and maintenance are emphasized while most watchstanding is stopped.

Often we say that trying to fix a problem by throwing money at it is an inadequate solution. With regard to crew fatigue, money can help quite a bit if it is used to make sure that adequate numbers of crewmembers are available to offset the need for sleep.

Supplemental melatonin use will probably not be practical for use while underway. It can produce increased sleep inertia. I expect that its effects would be overcome easily by adrenaline, but maybe not. A good alternative is to use protection from bright light and light therapy when appropriate. Perhaps melatonin would be useful during a port call, but the crew may be tired enough that supplemental melatonin would not be needed.

What Would I Do?

Having pondered sleep, circadian rhythms and fatigue for 40 years, shiftwork planning for 30+ years and watchstanding plans for about 20 years, I do have some recommendations. These fall into the categories of safety maxims, selection criteria, knowing your body clock, tactical caffeine use, and careful scheduling during port calls.

Two Safety Maxims

I have stated quite often over the years that requiring humans to work at night and sleep during the day is a "crime against nature." Our biology simply does not work that way. The results of meeting this requirement include:

> The circadian pattern of accidents with the astonishingly sharp peak in the pre-dawn hours and the secondary peak in the mid-afternoon.

> Trying to accomplish a safety-sensitive job (involving personal, worker and/or public risk) with less than eight hours of sleep in the preceding 24 hours and/or more than 17 hours of continuous wakefulness.

The demand for night work and day sleep causes accidents (197). Thus...

> **Safety maxim 1:** Mental fatigue will always be an accident risk factor when 24/7 operations are required.

> Generally, people place sleep and work at two ends of a spectrum of accomplishment. When safety is not issue, bosses still penalize workers for sleeping on the job. They do this even though a brief nap taken by a fatigued worker during a lull in operations might help prevent a work-related calamity later in the work period. High achievers tell us "I can sleep when I die." These are pretty stupid approaches to safety. Thus...

> **Safety maxim 2:** When 24/7 operations are required, good sleep hygiene is part of the job requirement.

Signing Up for Watchstanding

Some folks are better suited to meet the demands of 24/7 operations that others. Having knowledge of selection criteria can help individuals decide if they want to try night work, and can help employers decide who should or should not be on night work. I listed a number of contraindications that suggest that an individual may not be well-suited for shiftwork or night work. I also listed a number of indications that suggest that an individual is dealing poorly with the night work requirement. The use of this information by employers may lead to safer operations and healthier employees.

Know Your Body

Each watchstander should be aware of his or her chronotype: are you a "lark" or an "owl"? This knowledge can help a watchstander plan effective sleep periods. Each watchstander should learn how to conduct Autorhythmometry by Tympanic Thermometry (ATT). Difficult to pronounce, but pretty easy to do. The idea is to take your temperature every hour or two and write down (or graph) the reading. Oral temperature is OK as long as it is taken at least fifteen minutes after eating or drinking anything. Foods and liquids change the temperature of the mouth. Ear drum (tympanic membrane) temperature is a much better indication of body temperature than oral temperature. As a physiologist, I am absolutely not a fan of the infrared devices that measure skin temperature. These may indicate when a sick person has a fever, but they are useless for estimating your circadian rhythm. Here's how to estimate your circadian rhythm in body temperature.

First, buy an infrared thermometer that records "ear" temperature. They are available at pharmacies and other stores. Second, take your temperature properly. Hold the thermometer in one hand. Reach over your head with your free hand, grasp the top of your ear and tug up on the ear gently. This tug tends to straighten the ear canal, allowing the thermometer's infrared sensor to be aimed more directly at the ear drum and less at surrounding, cooler tissues. With the other hand, insert the thermometer, with a clean, disposable, plastic cover, into the ear canal. Aim the thermometer approximately at the back of the opposite eyeball. Rotate the thermometer about 1/4 turn to seat it, then press the sensing button and hold it down for at least the one second needed for a reading. Note the temperature, discard the dirty cover, acquire a new cover, and repeat the process to get a second reading. Write down the higher of the two readings. Taking more than two readings will cause unwanted cooling in the ear canal. Always use the same ear.

Finally, whichever way you choose to measure your temperature, look for the highest temperature reading of the 24-hour period. For most folks it will occur in the early evening. For larks on a regular schedule (no night work), it may occur in the afternoon. For an owl on a regular schedule, it may occur late in the evening. Watch across days of watchstanding for two kinds of change. One change may be a flattening of the height (amplitude) of your circadian rhythm. This change may be associated with feelings of *malaise*. Another change may be a shifting of the peak earlier or later in the day. These indicate a phase advance or phase delay, respectively, in the body clock with respect to the day-night cycle. This change may be associated with feeling sleepy at earlier of later times of day than usual.

When you have been sleeping at night for at east several days and will be beginning watchstanding, you may wish to use phase advance or delay tactics for up to three days before the demand for night work. These are accomplished with exposure to bright light and the ingestion of supplemental melatonin. Your tactics must be planned on the basis of knowledge about your chronotype, the status of your body clock, and knowledge about your upcoming watchstanding plan. A lark may wish to phase advance the body clock by (1) going to bed an hour earlier each night for up to three nights and also arising an hour earlier each morning; (2) taking 1 to 5 mg melatonin in the mid-afternoon of these three days, before the earlier sleep period; and (3) being exposed to bright light and/or sunlight earlier each morning. An owl may wish to phase delay the body clock by (1) going to bed an hour later each night for up to three nights and also arising an hour later each morning; and (2) being exposed to bright light and/or sunlight later each evening. Only experience with these methods and different fixed and rotating watchstanding plans will allow you to make good decisions about your personal use of these body clock phase advance and phase delay tactics.

Tactical Caffeine Use

As noted earlier, I recommend the planning of tactical caffeine use. Recall that caffeine is present in some prescription and over-the-counter medications. For tactical use, caffeine may be acquired in liquid (coffee, mainly), gum and pill forms.

Port Call Time Management

The following two ideas will not be practical in many operations. However, I include them in the hope that they may be used occasionally to enhance operational safety. The first idea is that of standing down from watchstanding to allow crew recovery. I observed this practice in the mid-1990s on U.S. Coast Guard medium endurance cutters (WMECs). Approximately every ten days the cutter would tie up to a dock for up to 24 hours, allowing watchstanding to cease (of course, they remained on alert and no crewmembers went ashore). The in-port watch schedule would commence. The crew would have a good meal and then sleep as much as possible. Obviously, there are commercial problems with the availability of dock space and even anchoring space within a crowded harbor that will hamper the usability of this practice, but it might be usable in some circumstances, if allowed by company policies and financial issues.

A corollary of this idea would apply to two-section submarine operations. If possible, I would set the boat down for 12 to 14 hours, feed the crew a good meal and initiate a mandatory sleep period of ten hours.

The second idea is the use of a modified watchstanding plan during on/offloading and departure from port. The thought, for example, of a first mate spending 36 continuous hours supervising the offloading of ballast and onloading of crude oil in Valdez is disquieting. If such a practice is used, then there must be a policy and a plan to prevent the sleep-deprived individual from standing watch during the first 12 to 24 hours while underway. That person's assignment should be to get adequate sleep so that they may resume their watch rotation. In the meantime, it would not be advisable to over-commit other watchstanders to take up the slack from the watchstander in the sack. Thus, an extra watchstander must be available when one watchstander becomes sleep deprived due to in-port operations.

In closing, I hope that you have, in fact, found this book to be the most complete reference work available concerning (1) the genesis and history of maritime watchstanding and (2) more than a half-century of research concerning different watchstanding plans. Perhaps my assessments of more than 35 watchstanding plans from operations and/or laboratories will be of use to you.

I hope to publish a second edition of this book within a year. If you have additional information about maritime watchstanding practices from ancient times through the 1800s, relevant information about the development of the measurement time, and/or other watchstanding research literature or FRMS guidance that I've missed please contact me and send it along for possible inclusion in the second edition.

References

1. Allen P, Wadsworth E, Smith A. The prevention and management of seafarers' fatigue: a review. *Int. Marit. Health.* 2007;58:1–4.

2. Allen P, Wadsworth E, Smith A. Seafarers' fatigue: a review of the recent literature. *Int. Marit. Health.* 2008;59(1-4):81–92.

3. Alluisi EA, Chiles WD, Hall TJ, Hawkes GR. Human Group Performance during Confinement. Lockheed Aircraft Corp, Marietta GA; 1963. Available from: http://www.dtic.mil/docs/citations/AD0426661

4. Anonymous. *Project Horizon - A Wake-Up Call.* Southampton U.K.: Warshash Maritime Academy, Southampton Solent University; 2012. Available from: http://www.warsashacademy.co.uk/about/our-schools/maritime-research-centre/horizon-project/horizon-project.aspx

5. Antle DM, Côté JN. Relationships between lower limb and trunk discomfort and vascular, muscular and kinetic outcomes during stationary standing work. *Gait Posture.* 2013 Apr;37(4):615–619.

6. Anund A, Fors C, Kecklund G, van Leeuwen W, Åkerstedt T. *Countermeasures for Fatigue in Transportation: A review of existing methods for drivers on road, rail, sea and in aviation.* Linköping, Sweden: Swedish National Road and Transport Research Institute (VTI); 2015. Available from: http://www.diva-portal.org/smash/record.jsf?pid=diva2%3A807456&dswid=64

7. Arendt J, Middleton B, Williams P, Francis G, Luke C. Sleep and circadian phase in a ship's crew. *J. Biol. Rhythms.* 2006 Jun;21(3):214–221.

8. Arulanandam S, Tsing GCC. Comparison of alertness levels in ship crew. An experiment on rotating versus fixed watch schedules. *Int. Marit. Health.* 2009;60(1-2):6–9.

9. Australian Maritime Safety Authority. *Fatigue Risk Management Plan: The Default Plan* (AMSA 406). 2013 Mar;

10. Babbitt BA NYstrom CO. *Questionnaire Construction Manual.* Westlake

Village CA: Essex Corp.; 1989. Available from: http://stinet.dtic.mil/oai/oai?&verb=getRecord&metadataPrefix=html&identifier=ADA212365

11. Baert P, Rabat A, Trousselard M, Van-Beers P, Coste O. *Cumulative effects on physiological, psychological and cognitive performance of a light deprivation and shift-work conditions during 70 days in a patrol of a nuclear submarine ballistic missile.* French Navy Health Service (CSS/FOST) and the Army Biomedical Research Institute (IRBA); 2010.

12. Baillie Reynolds PK. *The Vigiles of Imperial Rome.* Chicago IL: Ares Pub.; 1996.

13. Baker CC, Malone T, Krull RD. *Survey of Maritime Experiences in Reduced Workload and Staffing.* Report no. CG-D-02-00. Groton CT: U.S. Coast Guard Research and Development Center, 1999.

14. Baker CC, McCafferty DB. Accident database review of human element concerns: What do the results mean for classification? In: *Proc. Int Conf.* London: *Royal Institution of Naval Architects*; 2005.

15. Balkin TJ, Rupp TL, Wesensten NJ, Bliese PD. Paying Down the Sleep Debt: Realization of Benefits During Subsequent Sleep Restriction and Recovery. In: *26th Army Science Conference.* Orlando FL: 2008. Available from: http://www.stormingmedia.us/51/5175/A517505.html

16. Barnett M, Lützhöft M, Akerstedt T. *Fatigue Management Toolkit.* Warshash Maritime Academy, Southampton Solent University; 2012.

17. Barr L, Howarth H, Popkin S, Carroll RJ. *A review and evaluation of emerging driver fatigue detection measures and technologies.* Cambridge MA: National Transportation Systems Center, US Department of Transportation; 2005.

18. Bearden B. *The Bluejackets' Manual.* 21st ed. Annapolis MD: U.S. Naval Institute; 1990.

19. Beare AN, Bondi KR, Biersner RJ, Naitoh P. Work and rest on nuclear submarines. *Ergonomics.* 1981 Aug;24(8):593–610.

20. Beck M. The Sleepless Elite. *Wall Str. J.* 2011 Apr 5;

21. Betts J. *Time Restored: The Story of the Harrison Timekeepers and R.T.*

Gould, The Man who Knew (almost) Everything. National Maritime Museum & Oxford; 2006.

22. Blassingame SR. *Analysis Of Self-Reported Sleep Patterns In A Sample Of Us Navy Submariners Using Nonparametric Statistics.* Monterey CA: Naval Postgraduate School; 2001. Available from: http://www.dtic.mil/docs/citations/ADA397301

23. Boorstin DJ. *The Discoverers.* New York: Vintage Books; 1985.

24. Born J. Slow-wave sleep and the consolidation of long-term memory. *World J. Biol. Psychiatry.* 2010 Jun;11 Suppl 1:16–21.

25. Born J, Wilhelm I. System consolidation of memory during sleep. *Psychol. Res.* 2012 Mar;76(2):192–203.

26. Boulard R, Telle-Lamberton M, Vezina M. *Organisation Temporelle du Travail sur les Navires de la Garde Cotiere Canadienne: Impacts Physiologiques.* Transport Canada; 1989. Available from: http://www.worldcat.org/title/organisation-temporelle-du-travail-sur-les-navires-de-la-garde-cotiere-canadienne-impacts-physiologiques-et-psychologiques/oclc/61793417?referer=di&ht=edition

27. Brink S. Sleepless Society. *US News World Rep.* 2000 Oct 16;129(15):63–72.

28. Broadbent DE. *Decision and Stress.* St. Louis MO: Academic Press; 1971. Available from: http://www.abebooks.com/Decision-Stress-Broadbent-D.E-Academic-Press/1867778576/bd

29. Broughton RJ. SCN controlled circadian arousal and the afternoon "nap zone." *Sleep Res. Online.* 1998;1(4):166–178.

30. Broughton R, Mullington J. Circasemidian sleep propensity and the phase-amplitude maintenance model of human sleep/wake regulation. *J. Sleep Res.* 1992;1(2):93–98.

31. Brown ID. *Study into Hours of Work, Fatigue and Safety at Sea.* Cambridge UK: Medical Research Council; 1989.

32. Brown L, Schoutens AMC, Whitehurst G, Booker TJ, Davis T, Losinski S, et al. *The Effect of Blue Light Therapy on Flight Crew-Members Behavioral Alertness.* Rochester NY: Social Science Research Network; 2014. Available from:

http://papers.ssrn.com/abstract=2402409

33. Brown SAT. *Maritime Platform Sleep and Performance Study: Evaluating the SAFTE Model for Maritime Workplace Application*. Monterey CA: Naval Postgraduate School; 2012.

34. Buckner DN, Harabedian A, Mcgrath JJ. *A Study of Individual Differences in Vigilance Performance*. Los Angeles CA: Human Factors Research Inc.; 1960. Available from: http://www.dtic.mil/docs/citations/AD0231897

35. Bullinger EW. *The Witness of the Stars*. 1974 edition. Kregel Publications; 1893.

36. Bureau of Navigation, Department of the Navy. *Articles for the Government of the United States Navy*. Navy Dep. Libr. Nav. Hist. Herit. Command. 1930; Available from: http://www.history.navy.mil/faqs/faq59-7.htm

37. Burkhart K, Phelps JR. Amber lenses to block blue light and improve sleep: a randomized trial. *Chronobiol. Int.* 2009 Dec;26(8):1602–1612.

38. Burkholder R. Solving the problem of longitude. *Cooks Log*. 1983;6(4):222.

39. Caille EJ, Bassano JL. Biorhythm and watch rhythms: hemeral watch rhythm and anhemeral watch rhythm in simulated permanent duty. In: RR Mackie (ed.), *Vigilance: Theory, Operational Performance, and Physiological Correlates*. New York: Plenum Press; 1977. p. 461–510.

40. Caldwell JA, Caldwell JL. *Fatigue in Aviation: A Guide to Staying Awake at the Stick*. Ashgate Publishing; 2004.

41. Caldwell JA, Caldwell JL, Brown D, Smythe N, Smith J. *The Effects of 37 Hours of Continuous Wakefulness on the Physiological Arousal, Cognitive Performance, Self-Reported Mood, and Simulator Flight Performance of F-117A Pilots*. Brooks AFB TX: Air Force Research Laboratory; 2003. Available from: http://www.dtic.mil/srch/doc?collection=t3&id=ADA415792

42. Caldwell JA, Caldwell JL, Smith J, Alvarado L, Heintz T. *The Efficacy of Modafinil for Sustaining Alertness and Simulator Flight Performance in F-117 Pilots During 37 Hours of Continuous Wakefulness*. Brooks AFB TX: Air Force Research Laboratory; 2004. Available from: http://www.dtic.mil/srch/doc?collection=t3&id=ADA420330

43. Caldwell JA, Mallis MM, Caldwell JL, Paul MA, Miller JC, Neri DF. Fatigue countermeasures in aviation. *Aviat. Space Environ. Med.* 2009 Jan;80(1):29–59.

44. Caldwell JA, Prazinko B, Caldwell JL. Body posture affects electroencephalographic activity and psychomotor vigilance task performance in sleep-deprived subjects. *Clin. Neurophysiol.* 2003 Jan;114(1):23–31.

45. Caldwell JA, Prazinko BF, Hall KK. The effects of body posture on resting electroencephalographic activity in sleep-deprived subjects. *Clin. Neurophysiol.* 2000 Mar;111(3):464–470.

46. Caldwell JA, Wesensten NJ, (ed.). Biomonitoring for Physiological and Cognitive Performance during Military Operations. In: *Proceedings of the SPIE.* Orlando FL: 2005. Available from: http://adsabs.harvard.edu/abs/2005SPIE.5797.....C

47. Calhoun SR. *Human Factors in Ship Design: Preventing and Reducing Shipboard Operator Fatigue.* 2006 Jun 12. Available from: http://ardujenski.com/files/documents/FatigueDesign.pdf

48. Chaiken S. *A Verification and Analysis of the USAF/DoD Fatigue Model and Fatigue Management Technology.* Brooks City-Base TX: Air Force Research Laboratory; 2005. Available from: http://www.stormingmedia.us/54/5453/A545344.html

49. Chapman S. *Fatigue Management amongst Royal Australian Navy Submariners.* HMAS Stirling, Royal Australian Navy; 2001.

50. Chapman S. *The management of stress and fatigue amongst Royal Australian Navy submariners☐ : a strategic, operational and financial imperative.* Australian Defence Headquarters, Department of Defence, 2001.

51. Chiles W, Alluisi E, Adams O. Work schedules and performance during confinement. *Hum. Factors.* 1968 Apr;10(2):143–196.

52. Cho K, Ennaceur A, Cole JC, Suh CK. Chronic jet lag produces cognitive deficits. *J. Neurosci.* 2000 Mar 15;20(6):RC66.

53. Cole RJ, Kripke DF, Gruen W, Mullaney DJ, Gillin JC. Automatic sleep/wake identification from wrist activity. *Sleep.* 1992;15(5):461–469.

54. Colgate S. *Fundamentals of Sailing, Cruising, and Racing.* New York: W.

W. Norton & Company; 1996.

55. Collins A, Matthews V, McNamara R. *Fatigue, Health and Injury among Seafarers and Workers on Offshore Installations: A Review*. Cardiff University, Wales: Seafarers International Research Centre (SIRC); 2000. Available from: http://www.offshorecenter.dk/log/bibliotek/mcga-survey-project461_app1.pdf.pdf

56. Colquhoun W, Blake M, Edwards R. Experimental studies of shift-work I: A comparison of "rotating" and "stabilized" 4-hour shift systems. *Ergonomics*. 1968 Sep;11(5):437–453.

57. Colquhoun WP. Hours of work at sea - Watchkeeping schedules, circadian rhythms and efficiency. *Ergonomics*. 1985 Apr;28:637–653.

58. Colquhoun W, Paine M, Fort A. Circadian rhythm of body temperature during prolonged undersea voyages. *Aviat. Space Environ. Med.* 1978 May;49(5):671–678.

59. Colquhoun W, Paine M, Fort A. Changes in the temperature rhythm of submariners following a rapidly rotating watchkeeping system for a prolonged period. *Int. Arch. Occup. Environ. Health*. 1979 Sep;42(3-4):185–190.

60. Colquhoun WP, Blake MJF, Edwards BS. Experimental Studies of Shift-Work II: Stabilized 8-hour Shift Systems. *Ergonomics*. 1968 Nov;11(6):527–546.

61. Colquhoun WP, Blake MJF, Edwards RS. *Experimental Studies of Watchkeeping*. Cambridge: Royal Naval Personnel Research Committee, Medical Research Council; 1969. Available from: http://stinet.dtic.mil/oai/oai?&verb=getRecord&metadataPrefix=html&identifier=AD0734528

62. Colquhoun WP, Blake MJF, Edwards RS. Experimental studies of shift-work. 3. Stabilized 12-hour shift system. *Ergonomics*. 1969 Nov;12(6):865–882.

63. Colquhoun WP, Folkard S. Scheduling Watches at Sea. In: Folkard S, Monk TH (ed.), *Hours of Work*. Chichester: John Wiley & Sons; 1985. p. 253–261.

64. Colquhoun WP, Hamilton P, Edwards RS. Effects of circadian rhythm, sleep deprivation, and fatigue on watchkeeping

performance during the night hours. In: Colquhoun WP, Rutenfranz J, (ed.). *Studies of Shiftwork*. London: Taylor & Francis; 1980. p. 225–233.

65. Colquhoun WP, Rutenfranz J, Goethe H, Neidhart B, Condon R, Plett R, et al. Work at sea: a study of sleep, and of circadian rhythms in physiological and psychological functions, in watchkeepers on merchant vessels. I. Watchkeeping on board ships: a methodological approach. *Int. Arch. Occup. Environ. Health*. 1988;60(5):321–329.

66. Colquhoun WP, Watson KJ, Gordon DS. Shipboard study of a four-crew rotating watchkeeping system. *Ergonomics*. 1987;30(9):1341–1352.

67. Committee on Military Nutrition Research, Food and Nutrition Board. *Caffeine for the Sustainment of Mental Task Performance: Formulations for Military Operations*. Washington, D.C.: The National Academies Press; 2001.

68. Comperatore CA, Caldwell JA, Caldwell JL. *Leader's Guide to Crew Endurance*. Fort Rucker AL: U.S. Army Aeromedical Research Laboratory and U.S Army Safety Center; 1997.

69. Comperatore CA, Kingsley L, Kirby AW, Rivera PK. *Management of Endurance Risk Factors: A Guide for Deep Draft Vessels*. 2001. Available from: http://stinet.dtic.mil/oai/oai?&verb=getRecord&metadataPrefix=html&identifier=ADA394262

70. Comperatore CA, Kirby A, Bloch C, Ferry C. *Alertness Degradation and Circadian Disruption on a U.S. Coast Guard Cutter Under Paragon Crewing Limits*. Groton CT: Coast Guard Research and Development Center; 1999. Available from: http://www.dtic.mil/srch/doc?collection=t3&id=ADA371228

71. Comperatore CA, Rivera PK. *Crew Endurance Management Practices: A Guide for Maritime Operations*. Groton CT: Coast Guard Research and Development Center. 2003. Available from: http://www.dtic.mil/docs/citations/ADA413881

72. Comperatore CA, Rivera PK, Carvalhais AB. *U.S. Coast Guard Guide for Managing Crew Endurance Risk Factors*. Version 2.1. Groton CT:

Coast Guard Research and Development Center; 2005. Available from: http://www.uscg.mil/SAFETY/docs/CEM/CEM_Guide_v2.pdf

73. Comperatore CA, Rothblum AM, Rivera PK, Kingsley LC, Beene D. *U.S. Coast Guard Guide for the Management of Crew Endurance Risk Factors*. Version 1.1. Groton CT: Coast Guard Research and Development Center; 2001. Available from: http://www.dtic.mil/docs/citations/ADA396127

74. Comperatore CA, Rothblum AM, Rivera PK, Kingsley LC, Beene D, Carvalhais AB. *U.S. Coast Guard Guide for the Management of Crew Endurance Risk Factors*. Version 1.0. Groton CT: Coast Guard Research and Development Center; 2001. Available from: http://www.uscg.mil/hq/cg9/rdc/reports/2001/CGD1301Report.pdf

75. Comperatore C, Bloch C, Ferry C. *Incidence of Sleep Loss and Wakefulness Degradation on a U.S. Coast Guard Cutter under Exemplar Crewing Limits*. Report no. CG-D-14-99. Groton CT: U.S. Coast Guard Research and Development Center, 1999.

76. Condon R, Colquhoun WP, Plett R, Knauth P, Fletcher N, Eickhoff S, et al. Circadian variation in performance and alertness under different work routines on ships. In: Haider M, Cervinka R (ed.). *Night and Shift Work: Long-term Effects and their Prevention*. Frankfurt: Peter Lang Publishing; 1986. p. 277–284.

77. Condon R, Knauth P, Klimmer F, Colquhoun WP, Herrmann H, Rutenfranz J. Adjustment of the oral temperature rhythm to a fixed watchkeeping system on board ship. *Int. Arch. Occup. Environ. Health*. 1984;54(2):173–180.

78. Cordle J, Shattuck NL. *A sea change in standing watch*. U. S. Nav. Inst. Proc. 2013 Jan;139(1):34–39.

79. Cowan HJ. *Time and Its Measurement: From The Stone Age To The Nuclear Age*. Cleveland OH: World Publishing Co.; 1958.

80. Craig A, Condon R. Operational Efficiency and Time of Day. *Hum. Factors*. 1984 Apr 1;26(2):197–205.

81. Crepeau L, Steele C, Duplessis C. At-sea evaluation of an alternative submarine watchstanding schedule. Undersea and Hyperbaric

Medical Society (UHMS) Annual Meeting, Oakland CA, 15 Sep 2006. Available from: http://archive.rubicon-foundation.org/xmlui/handle/123456789/3763

82. Crowley S, Acebo C, Carskadon M. Sleep, circadian rhythms, and delayed phase in adolescence. *Sleep Med.* 2007;8(6):602–612.

83. Crowley S, Lee C, Tseng C, Fogg L, Eastman C. Combinations of bright light, scheduled dark, sunglasses, and melatonin to facilitate circadian entrainment to night shift work. *J. Biol. Rhythms.* 2003 Dec;18(6):513–523.

84. Cutler TJ. *The Bluejackets' Manual.* 23rd (Centennial) Edition. Annapolis MD: U.S. Naval Institute; 2002.

85. Dahlgren A, van Leeuwen W, Kircher A, Lützhöft M, Kecklund G, Barnett M, et al. Fatigue and sleepiness in seafarers working in a 6-on 6-off shift system – results from one week of simulated navigation work. 21st International Symposium on Shift work and Working Time: Biological mechanisms, recovery and risk management in the 24 h Society, Stockholm, 28 Jul 2011.

86. D'Amico AD, Kaufman E, Saxe C. *A Simulation Study of the Effects of Sleep Deprivation, Time of Watch, and Length of Time on Watch on Watchstanding Effectiveness.* Kings Point NY: Computer Aided Operations Research Facility, National Maritime Research Center 1986. Available from: http://www.dtic.mil/docs/citations/ADA167729

87. Daniel C, Lamb J. NSMRL: *A Small Command with a Huge Presence for the Submarine Force.* Groton CT: Naval Submarine Medical Research Laboratory; 2005.

88. Davey JD. *Effects of Sleep Deprivation on U.S. Navy Watchstander Performance Onboard the Independence Class Littoral Combat Ship (LCS 2).* Monterey CA: Naval Postgraduate School; 2013. Available from: http://www.dtic.mil/docs/citations/ADA589928

89. Davis DL. *Commercial Navigation in the Greek and Roman World.* Doctoral Dissertation. Austin TX: University of Texas, 2009;

90. Davis SC, Cameron BJ, Helsegrave RJ. *Study on Extended Coast Guard Crewing Periods, Phase 3.* Montreal, Quebec: Transport Canada; 1999.

91. Davis SC Cameron BJ, Helsegrave RJ, Hamilton K. *Study on Extended*

Coast Guard Crewing Periods, Phase 2. Montreal, Quebec: Transport Canada; 1998.

92. Davis SC, Hamilton K, Heslegrave RJ Cameron BJ. *Study on Extended Coast Guard Crewing Periods*. Montreal, Quebec: Transport Canada; 1997.

93. Dinges DF, Pack F, Williams K, Gillen KA, Powell JW, Ott GE, et al. Cumulative sleepiness, mood disturbance, and psychomotor vigilance performance decrements during a week of sleep restricted to 4-5 hours per night. *Sleep*. 1997 Apr;20(4):267–277.

94. Dinges DF, Powell JW. Microcomputer analyses of performance on a portable simple visual RT task during sustained operations. *Behav. Res. Methods Instrum. Comput.* 1985;17(6):652–655.

95. Dingus TA, Hardee HL, Wierwille WW. Development of models for on-board detection of driver impairment. *Accid. Anal. Prev.* 1987 Aug;19(4):271–283.

96. Donderi DC, Smiley A, Kawaja KM. Shift Schedule Comparison for the Canadian Coast Guard. Transport Canada; 1995.

97. Duff IF. Medical aspects of submarine Warfare: the human factor as reflected in war patrol reports. In: Shilling CW, Kohl JW, (ed.). *History of Submarine Medicine in World War II* (Report no. 112). New London CT: U.S. Naval Submarine Base; 1947. p. 99–116. Available from: https://ia801702.us.archive.org/33/items/SubmarineMedicineInWorldWarII/Submarine%20Medicine%20in%20World%20War%20II.pdf

98. Duff IF, Shilling CW. Psychiatric casualties in submarine warfare. In: Shilling CW, Kohl JW, (ed.). *History of Submarine Medicine in World War II* (Report no. 112). New London CT: U.S. Naval Submarine Base; 1947. p. 117–128. Available from: https://ia801702.us.archive.org/33/items/SubmarineMedicineInWorldWarII/Submarine%20Medicine%20in%20World%20War%20II.pdf

99. Duffy JF, Czeisler CA. Effect of Light on Human Circadian Physiology. *Sleep Med. Clin.* 2009 Jun;4(2):165–177.

100. Duplessis CA, Miller JC, Crepeau LJ, Osborn CM, Dyche J.

Submarine watch schedules: underway evaluation of rotating (contemporary) and compressed (alternative) schedules. *Undersea Hyperb. Med.* 2007 Feb;34(1):21–33.

101. Eastman CI. Are separate temperature and activity oscillators necessary to explain the phenomena of human circadian rhythms? In: Moore-Ede MC, Czeisler CA (ed.). *Mathematical models of the circadian sleep-wake cycle.* New York: Raven Press; 1984. p. 81–103.

102. Ekirch AR. Sleep we have lost: pre-industrial slumber in the British Isles. *Am. Hist. Rev.* 2001;106(2):343–386.

103. Ekirch AR. *At Day's Close: Night in Times Past.* W. W. Norton & Company; 2006.

104. Engelbrecht EA (ed.). *The Lutheran Study Bible.* St Louis MO: Concordia Publishing House; 2009.

105. Engelbrecht EA. *The Church from Age to Age: A History from Galilee to Global Christianity.* St Louis MO: Concordia Publishing House; 2011.

106. Eriksen CA, Gillberg M, Vestergren P. Sleepiness and sleep in a simulated "six hours on/six hours off" sea watch system. *Chronobiol. Int.* 2006;23(6):1193–1202.

107. Eskandarian A, Sayed R, Delaigue P, Mortazavi A, Blum J. *Advanced Driver Fatigue Research.* 2007 Apr. Available from: http://trid.trb.org/view.aspx?id=898430

108. Fletcher N, Colquhoun WP, Knauth P, De Vol D, Plett R. Work at sea: a study of sleep, and of circadian rhythms in physiological and psychological functions, in watchkeepers on merchant vessels. VI. A sea trial of an alternative watchkeeping system for the merchant marine. *Int. Arch. Occup. Environ. Health.* 1988;61(1-2):51–57.

109. Folkard S, Tucker P. Shift work, safety and productivity. *Occup. Med. Oxf. Engl.* 2003 Mar;53(2):95–101.

110. Freeth T, Bitsakis Y, Moussas X, Seiradakis JH, Tselikas A, Mangou H, et al. Decoding the ancient Greek astronomical calculator known as the Antikythera Mechanism. *Nature.* 2006 Nov 30;444(7119):587–591.

111. Freeth T, Jones A, Steele JM, Bitsakis Y. Calendars with Olympiad

       display and eclipse prediction on the Antikythera Mechanism. *Nature.* 2008 Jul 1;454:614–617.

112. Fronto MC. The Roman Army AD 163. In: Marcus Cornelius Fronto: Correspondence. London: 1920. p. 147–151, 207–213.Available from: http://www.angelfire.com/tv2/shows4/RMil.html

113. Fuentes G, Lombardo T. A better watchbill. *Navy Times.* 2012 Jun 25;20–22.

114. Gander PH. *A Review of Fatigue Management in the Maritime Sector.* Massey University: Sleep Wake Research Centre; 2005.

115. Gander PH, van den Berg M, Signal L. Sleep and sleepiness of fishermen on rotating schedules. *Chronobiol. Int.* 2008 Apr;25(2):389–398.

116. Garfield B. *The thousand-mile war World War II in Alaska and the Aleutians.* Fairbanks: University of Alaska Press; 1995.

117. Gilliland K, Schlegel RE. *Readiness to Perform Testing: A Critical Analysis of the Concept and Current Practices.* Norman OK: University of Oklahoma; 1993. Available from: http://stinet.dtic.mil/oai/oai?&verb=getRecord&metadataPrefix=html&identifier=ADA269379

118. Golz PDM, Sommer D, Trutschel U, Sirois B, Edwards D. Evaluation of fatigue monitoring technologies. *Somnologie - Schlafforschung Schlafmed.* 2010 Sep 1;14(3):187–199.

119. Green DM, Swets JA. *Signal Detection Theory and Psychophysics.* Oxford, England: Robert E. Krieger; 1974.

120. Green KY. *A Comparative Analysis Between the Navy Standard Workweek and the Actual Work/Rest Patterns of Sailors Aboard U.S. Navy Frigates.* Monterey CA: Naval Postgraduate School; 2009. Available from: http://www.dtic.mil/docs/citations/ADA514116

121. Halberg F, Johnson EA, Nelson W, Runge W, Sothern E. Autorhythmometry procedures for physiologic self-measurements and their analysis. *Physiol. Teach.* 1972;1:1–11.

122. Härmä MI, Partinen M, Repo R, Sorsa M, Siivonen P. Effects of 6/6 and 4/8 watch systems on sleepiness among bridge officers.

Chronobiol. Int. 2008 Apr;25(2):413–423.

123. Harrison Y, Horne JA. The impact of sleep deprivation on decision making: a review. *J. Exp. Psychol. Appl.* 2000 Sep;6(3):236–249.

124. Harris W, O'Hanlon JF. *A Study of Recovery Funtions in Man.* Goleta Ca: Human Factors Research Div, Canyon Research Group Inc.; 1972. Available from: http://www.dtic.mil/docs/citations/AD0741828

125. Hartley L, Horberry T, Mabbott N, Kreuger G. *Review of Fatigue Detection and Prediction Technologies.* Australia: National Road Transport Commission; 2000. Available from: file:///C:/Documents%20and%20Settings/James%20Miller/My%20Documents/Miller%20Ergonomics/HTML%20Projects/FatiguéRef/biblio.htm#hartley00

126. Hartman BO, Cantrell GK. MOL: *Crew Performance on Demanding Work/Rest Schedules Compounded by Sleep Deprivation.* Brooks AFB TX: USAF School of Aerospace Medicine; 1967. Available from: http://stinet.dtic.mil/oai/oai?&verb=getRecord&metadataPrefix=html&identifier=AD0665845

127. al-Hassan AY (ed.). *Science and Technology in Islam, Part I: The Exact and Natural Sciences.* UNESCO Publishing; 2002. Available from: http://publishing.unesco.org/details.aspx?&Code_Livre=3424&change=E

128. Haswell JE. *Horology.* Facsimile of 1947 Edition. Wakefield: Charles River Books; 1976.

129. Hayashi M, Morikawa T, Hori T. Circasemidian 12 h cycle of slow wave sleep under constant darkness. *Clin. Neurophysiol.* 2002 Sep;113(9):1505–1516.

130. Haynes LE. *A Comparison Between the Navy Standard Workweek and Actual Work and Rest Patterns of U.S. Navy Sailors.* Monterey CA: Naval Postgraduate School; 2007. Available from: http://www.dtic.mil/docs/citations/ADA474039

131. Heslegrave RJ, Davis S, Cameron BJ. Fatigue Management: *A Guide for Canadian Coast Guard Managers, Officers and Crew.* Fisheries and Oceans Canada and Coast Guard; 1999. Available from: http://publications.gc.ca/site/eng/86273/publication.html

132. Hoddes E, Zarcone V, Smythe H, Phillips R, Dement WC. Quantification of sleepiness: a new approach. *Psychophysiology*. 1973 Jul;10(4):431–436.

133. Horne JA, Ostberg O. A self-assessment questionnaire to determine morningness-eveningness in human circadian rhythms. *Int. J. Chronobiol.* 1976;4(2):97–110.

134. Howarth HD, Pratt JH, Tepas DI. Do maritime crew members have sleep disturbances? *Int. J. Occup. Environ. Health.* 1999 Jun;5(2):95–100.

135. Hunt P, Kelly T. *Light Levels Aboard a Submarine: Results of a Survey with a Discussion of the Implications for Circadian Rhythms.* San Diego CA: Naval Health Research Center; 1995. Available from: http://www.ntis.gov/search/product.aspx?ABBR=ADA298064

136. Hursh SR, Balkin TJ. Response to commentary on fatigue models for applied research in warfighting. *Aviat. Space Environ. Med.* 2004 Mar;75(3 Section II):A57–A60.

137. Hursh SR, Balkin TJ, Miller JC, Eddy DR. The fatigue avoidance scheduling tool: Modeling to minimize the effects of fatigue on cognitive performance. *SAE Trans.* 2004;113(1):111–119.

138. Hursh SR, Eddy DR, Elsmore T. The Fatigue Avoidance Scheduling Tool: Modeling to Minimize the Effects of Fatigue on Cognitive Performance. Digital Human Modeling for Design and Engineering Symposium, Rochester MI, Jun 2004. Available from: https://www.sae.org/technical/papers/2004-01-2151

139. Hursh SR, Raslear TG, Kaye AS, Fanzone JF. *Validation and Calibration of a Fatigue Assessment Tool for Railroad Work Schedules.* Washington DC: Transportation Research Bord; 2008. Available from: http://tris.trb.org/view.aspx?type=MO&id=915759

140. Hursh SR, Redmond DP, Johnson ML, Thorne DR, Belenky G, Balkin TJ, et al. Fatigue models for applied research in warfighting. *Aviat. Space Environ. Med.* 2004 Mar;75(3 Suppl):A44–53; discussion A54–60.

141. Hursh SR, Redmond DP, Johnson ML, Thorne DR, Belenky G, Balkin TJ, et al. The DOD Sleep, Activity, Fatigue, and Task Effectiveness Model (03-BRIMS-001). 2003 May. Available from:

http://www.sisostds.org/index.php?tg=articles&idx=More&topics=71&article=229

142. Jacobsen BS, Thoft E, Grontmij C. *Shipping and Rest: How We Can Do Better*. Seahealth Denmark, Copenhagen, 2010.

143. Jay SM, Lamond M, Ferguson SA, Dorrian J, Jones CB, Dawson D. The characteristics of recovery sleep when recovery opportunity is restricted. *Sleep*. 2007;30(3):353–360.

144. Jean GV. Duty Aboard the Littoral Combat Ship: "Grueling but Manageable". *Natl. Def.* 2010 Sep. Available from: http://www.nationaldefensemagazine.org/archive/2010/September/Pages/DutyAboardtheLittoralCombatShip"GruelingbutManageable".aspx

145. Johnson LC, Naitoh P. *The Operational Consequences of Sleep Deprivation and Sleep Deficit*. Neuilly sur Seine, France: Advisory Group for Aerospace Research and Development; 1974. Available from: http://oai.dtic.mil/oai/oai?verb=getRecord&metadataPrefix=html&identifier=AD0783199

146. Jores A. Physiologie und Pathologie der 24-Stunden-Rhythmik des Menschen. In: Czerny A, Kraus F, Müller F, Pfaundler M v, Schittenhelm A, (ed.). *Ergebnisse der Inneren Medizin und Kinderheilkunde*. Springer Berlin Heidelberg; 1935. p. 574–629.Available from: http://link.springer.com/chapter/10.1007/978-3-642-90670-1_11

147. Josephus F. *Josephus: The Essential Writings*. Grand Rapids, MI: Kregel; 1994.

148. Kaida K, Takahashi M, Akerstedt T, Nakata A, Otsuka Y, Haratani T, et al. Validation of the Karolinska sleepiness scale against performance and EEG variables. *Clin. Neurophysiol.* 2006 Jul;117(7):1574–1581.

149. Kelly T, Grill J, Ryman D, Hunt P, Dijk D, Mitchell J, et al. *Circadian Rhythms in Submariners Scheduled to an 18-Hour Day*. San Diego CA: Naval Health Research Center; 1996.

150. Kelly TL, Grill JT, Hunt PD, Neri D F. *Submarines and 18-Hour Work Shift Work Schedules*. San Diego CA: Naval Health Research Center; 1996. Available from:

http://www.dtic.mil/docs/citations/ADA306497

151. Kelly T, Neri D, Grill J, Ryman D, Hunt P, Dijk D, et al. Nonentrained Circadian Rhythms of Melatonin in Submariners Scheduled to an 18-Hour Day. *J Biol Rhythms*. 1999 Jun 1;14(3):190–196.

152. Kelly T, Ryman D, Pattison S. *A Comparison of Two Navy Watch Schedules*. San Diego CA: Naval Health Research Center; 1996. Available from: http://www.dtic.mil/srch/doc?collection=t3&id=ADA323196

153. Keppie L. *The Making of the Roman Army: From Republic to Empire*. University of Oklahoma Press; 1998.

154. Kinnaman SA (ed.). *Treasury of Daily Prayer*. St. Louis, MO: Concordia Pub. House; 2008.

155. Kleitman N. *Sleep and Wakefulness*. 2nd ed. Chicago IL: University of Chicago Press; 1963.

156. Kleitman N. The sleep-wakefulness cycle of submarine personnel. In: *Human Factors in Undersea Warfare*. Baltimore: Waverly Press; 1949. p. 329–341.

157. Kleitman N, Jackson D. *Variations in Body Temperature and in Performance Under Different Watch Schedules*. Naval Medical Research Institute, Bethesda MD; 1950. Available from: http://www.dtic.mil/dtic/tr/fulltext/u2/667515.pdf

158. Knauth P, Condon R, Klimmer F, Colquhoun WP, Hermann H, Rutenfranz J. Sleep data sampled from the crew of a merchant marine ship. In: *Breakdown in Human Adaptation to "Stress."* Hingham MA: Nijhoff Publishers; 1984.

159. Knauth P, Landau K, Droge C, Schitteck M, Widynski M, Rutenfranz J. Duration of sleep depending on the type of shift work. *Int. Arch. Occup. Environ. Health*. 1980;46(2):167–177.

160. Knauth P, Rohmert W, Rutenfranz J. Systematic selection of shift plans for continuous production with the aid of work-physiological criteria. *Appl. Ergon.* 1979 Mar;10(1):9–15.

161. Knauth P, Rutenfranz J. Experimental shift work studies of permanent night, and rapidly rotating, shift systems I. *Int. Arch.*

Occup. Environ. Health. 1976;37(2):125–137.

162. Lamb JC, Huebner HM, Fitzgerald MP. *Naval Submarine Medical Research Laboratory Command History, OPNAV 5750-1 Fiscal Year 2005*. Groton CT: Naval Submarine Medical Research Laboratory; 2006. Available from: http://www.dtic.mil/docs/citations/ADA457076

163. Laperrière E, Ngomo S, Thibault M-C, Messing K. Indicators for choosing an optimal mix of major working postures. *Appl. Ergon.* 2006 May;37(3):349–357.

164. Lauer JT. *Sleep Deprivation and the Surface Navy: Possible Impacts on the Performance of Command and Control Personnel*. Monterey CA: Naval Postgraduate School; 1991.

165. Lavie P, Segal S. Twenty-four-hour structure of sleepiness in morning and evening persons investigated by ultrashort sleep-wake cycle. *Sleep*. 1989 Dec;12(6):522–528.

166. Lee JD, McCallum MC, Maloney AL, Jamieson GA. *Validation and Sensitivity Analysis of a Crew Size Evaluation Method*. Seattle WA: Battelle Research Center; 1997.

167. Lee JD, Sanquist TF. *Human Factors Plan for Maritime Safety: Annotated Bibliography*. Seattle WA: Battelle Human Affairs Research Centers Seattle Wa; 1993. Available from: http://www.dtic.mil/docs/citations/ADA265392

168. Van Leeuwen WMA, Kircher A, Dahlgren A, Lützhöft M, Barnett M, Kecklund G, et al. Sleep, sleepiness, and neurobehavioral performance while on watch in a simulated 4 hours on/8 hours off maritime watch system. *Chronobiol. Int.* 2013 Nov;30(9):1108–1115.

169. Van Leeuwen W, Dahlgren A, Kircher A, Lützhöft M, Barnett M, Kecklund G, et al. *Comparing subjective and objective sleepiness between the two most common maritime watch systems: a bridge simulator study*. 21st International Symposium on Shift work and Working Time: Biological mechanisms, recovery and risk management in the 24 h Society, Stockholm, 28 Jul 2011.

170. Van Leeuwen W, Dahlgren A, Kircher A, Lützhöft M, Barnett M, Kecklund G, et al. *Subjective and objective sleepiness in a simulated "4 hours on/8 hours off" maritime watch system*. 21st International

Symposium on Shift work and Working Time: Biological mechanisms, recovery and risk management in the 24 h Society, Stockholm, 28 Jul 2011.

171. Van der Lely S, Frey S, Garbazza C, Wirz-Justice A, Jenni OG, Steiner R, et al. Blue blocker glasses as a countermeasure for alerting effects of evening light-emitting diode screen exposure in male teenagers. *J. Adolesc. Health.* 2015 Jan;56(1):113–119.

172. Lewy AJ, Sack RL. The dim light melatonin onset as a marker for circadian phase position. *Chronobiol. Int.* 1989;6(1):93–102.

173. Lim J, Dinges DF. Sleep deprivation and vigilant attention. *Ann. N. Y. Acad. Sci.* 2008;1129:305–322.

174. Lundy B, Miller JC, Jackson K, Senchina DS, Burke LM, Stear SJ, et al. A–Z of nutritional supplements: dietary supplements, sports nutrition foods and ergogenic aids for health and performance – Part 25. *Br. J. Sports Med.* 2011 Oct 1;45(13):1077 –1078.

175. Lützhöft M, Dahlgren A, Kircher A, Thorslund B, Gillberg M. Fatigue at sea in Swedish shipping-a field study. *Am. J. Ind. Med.* 2010 Jul;53(7):733–740.

176. Mackie RR (ed.). *Vigilance: Theory, Operational Performance, and Physiological Correlates.* 1st ed. Springer; 1977.

177. Mackie RR. *Research on Factors Influencing the Interpretation of Sonar Signals.* Goleta CA: Human Factors Research Div, Canyon Research Group Inc. 1974. Available from: http://www.dtic.mil/docs/citations/ADA002307

178. Mackie RR, Miller JC. *Effects of Hours of Service, Regularity of Schedules and Cargo Loading on Truck and Bus Driver Fatigue.* Washington DC: National Highway Traffic Safety Administration; 1978. Available from: http://www.ntis.gov/search/product.aspx?ABBR=PB290957

179. Mackworth NH. The breakdown of vigilance durning prolonged visual search. *Q. J. Exp. Psychol.* 1948;1(1):6.

180. Madden PJ. *Jesus' Walking on the Sea: An Investigation of the Origin of the Narrative Account.* Berlin ; New York: Walter De Gruyter Inc; 1997.

181. Makeig S, Neri DF. *A Proposal for Integrated Shipboard Alertness Management*. San Diego CA: Naval Health Research Center. 1996. Available from: http://www.dtic.mil/docs/citations/ADA378421

182. Marine Accident Investigation Branch (MAIB). *Bridge Watchkeeping Safety Study*. Southampton U.K.: U.K. Department for Transport; 2004. Available from: http://www.maib.gov.uk/cms_resources.cfm?file=/Bridge_watch keeping_safety_study.pdf

183. Maritime Safety Committee. *Guidance on Fatigue Mitigation and Management*. London: United Nations International Maritime Organization (IMO); 2001. Available from: http://www.imo.org/OurWork/HumanElement/VisionPrinciples Goals/Documents/1014.pdf

184. Martin TW. Watch during the watches (Mark 13:35). *J. Biblic. Lit.* 2001;120(4):685–701.

185. Mason D. *A Comparative Analysis between the Navy Standard Workweek and the Work/Rest Patterns of Sailors Aboard U.S. Navy Cruisers*. Monterey CA: Naval Postgraduate School; 2009. Available from: http://faculty.nps.edu/nlmiller/millerth.htm

186. Mathew KM. *History of the Portuguese Navigation in India, 1497-1600*. Mittal Publications; 1988.

187. McCallum MC, Raby M, Rothblum AM. *Procedures for Investigating and Reporting Fatigue Contributions to Marine Casualties*. Seattle WA: Battelle Research Center; 1996. Available from: http://pro.sagepub.com/content/41/2/988

188. McCallum M, Sanquist T, Mitler M, Krueger G. *Commercial Transportation Operator Fatigue Management Reference*. Washington DC: U.S. Department of Transportation; 2003. Available from: hfcc.dot.gov/ofm/docs/fmr07-03.doc

189. McCauley ME, Royal JW, Wylie CD, O'Hanlon JF, Mackie RR. *Motion Sickness Incidence: Exploratory Studies of Habituation, Pitch and Roll, and the Refinement of a Mathematical Model*. Goleta CA: Human Factors Research Inc.; 1976. Available from: http://stinet.dtic.mil/oai/oai?&verb=getRecord&metadataPrefix= html&identifier=ADA024709

190. McClernon CK, Miller JC. Variance as a Measure of performance in an aviation context. *Int. J. Aviat. Psychol.* 2011;21(4):397–412.

191. McGrath JJ. *Subjective Reactions of Vigilance Performers.* Los Angeles CA: Human Factors Research Inc.; 1960. Available from: http://www.dtic.mil/docs/citations/AD0403026

192. McGrath JJ, Harabedian A, Buckner DN. *Review and critique of the literature on vigilance performance.* Los Angeles CA: Human Factors Research Inc.; 1959. Available from: http://oai.dtic.mil/oai/oai?verb=getRecord&metadataPrefix=html&identifier=AD0237691

193. Mcgrath JJ, Harabedian A, Buckner DN. *An Exploratory Study of the Correlates of Vigilance Performance.* Los Angeles CA: Human Factors Research Inc.; 1960. Available from: http://www.dtic.mil/docs/citations/AD0234087

194. Melfi ML, Miller JC. *Causes and Effects of Fatigue in Experienced Military Aircrew.* Brooks City-Base TX: Air Force Research Laboratory; 2006. Available from: http://www.dtic.mil/srch/doc?collection=t3&id=ADA462989

195. Miller D. *U-Boats: The Illustrated History of The Raiders Of The Deep.* Dulles, VA: Brassey's; 2002.

196. Miller JC. *Fundamentals of Shiftwork Scheduling, 3rd Edition: Fixing Stupid.* Smashwords; 2013. Available from: http://www.smashwords.com/books/view/352352

197. Miller JC. *Anatomy of a Fatigue-Related Accident.* Smashwords; 2013. Available from: https://www.smashwords.com/books/view/379949

198. Miller JC. *Usability Improvement for Data Input into the Fatigue Avoidance Scheduling Tool (FAST).* Brooks City-Base TX: Air Force Research Laboratory; 2005. Available from: http://www.dtic.mil/srch/doc?collection=t3&id=ADA435739

199. Miller JC. An historical view of operator fatigue. In: Matthews G, Desmond PA, Neubauer C, Hancock PA (ed.). *The Handbook of Operator Fatigue.* Surrey, England: Ashgate; 2012. p. 25–44. Available from: http://www.ashgate.com/isbn/9781409487005

200. Miller JC. *Shiftwork: An Annotated Bibliography.* Smashwords; 2013.

Available from: https://www.smashwords.com/books/view/300411

201. Miller JC. *Cognitive Performance Research at Brooks Air Force Base, Texas, 1960-2009*. Smashwords; 2013. Available from: http://www.amazon.com/Cognitive-Performance-Research-1960-2009-ebook/dp/B00C4CVIV2/ref=sr_1_1?s=books&ie=UTF8&qid=1365948764&sr=1-1&keywords=cognitive+performance+at+brooks+air

202. Miller JC. *In Search of Circasemidian Rhythms*. Brooks City-Base TX: Air Force Research Laboratory; 2006. Available from: http://stinet.dtic.mil/oai/oai?&verb=getRecord&metadataPrefix=html&identifier=ADA458153

203. Miller JC. *Sleep at Altitude*. 1976. Available from: http://proquest.umi.com/pqdweb?index=0&did=759905741&SrchMode=2&sid=3&Fmt=1&VInst=PROD&VType=PQD&RQT=309&VName=PQD&TS=1337856648&clientId=75697

204. Miller JC. Fit for Duty? *Ergon. Des.* 1996 Apr;4(2):11–17.

205. Miller JC. *21 Tips For Beating Fatigue And Improving Your Health, Happiness And Safety* (21 Book Series). First edition. CreateSpace Independent Publishing Platform; 2013.

206. Miller JC. *Fatigue Effects and Countermeasures in 24/7 Security Operations*. Alexandria VA: ASIS International; 2010.

207. Miller JC, Dyche J, Cardenas RJ, Carr W. *Effects of Three Watchstanding Schedules on Submariner Physiology, Performance and Mood*. Groton CT: Naval Submarine Medical Research Laboratory; 2003. Available from: http://www.ntis.gov/search/index.aspx

208. Miller JC, Eddy DR, Fischer J. *The Sensitivity and Specificity of Oculometrics Under Fatigue Stress Compared to Performance and Subjective Measures*. NTI Inc., Brooks City-Base TX; 2004. Available from: http://www.dtic.mil/docs/citations/ADA425455

209. Miller JC, Horvath SM. Cardiac output during sleep at altitude. *Aviat. Space Environ. Med.* 1977 Jul;48(7):621–624.

210. Miller JC, Lebegue BJ, Long JS, Pinchak AM, Herrera M. *The Effects of Three Lighting Conditions on Performance and Alertness in an Air Defense*

Operations Center (limited distribution). Brooks City-Base TX: Air Force Research Laboratory; 2008.

211. Miller JC, Mackie RR. *Vigilance Research and Nuclear Security: Critical Review and Potential Applications to Security Guard Performance*. Goleta CA: Human Factors Research Inc.; 1980.

212. Miller JC, Mitler MM. Predicting accident times. *Ergon. Des.* 1997;5(4):13–18.

213. Miller JC, Smith ML, McCauley ME. *Crew Fatigue and Performance on U.S. Coast Guard Cutters*. Groton CT: Coast Guard Research and Development Center; 1998. Available from: http://stinet.dtic.mil/oai/oai?&verb=getRecord&metadataPrefix=html&identifier=ADA366708

214. Miller NL, Firehammer R. Avoiding a Second Hollow Force: The Case for Including Crew Endurance Factors in the Afloat Staffing Policies of the US Navy. *Nav. Eng. J.* 2007;119(1):83–96.

215. Miller NL, Nguyen JL, Sanchez S, Miller JC. *Sleep Patterns and Fatigue Among U.S. Navy Sailors: Working the Night Shift During Combat Operations Aboard the USS STENNIS During Operation Enduring Freedom*. 2003 May. Available from: http://faculty.nps.edu/nlmiller/docs/AsMAUSSSTENNISAbstract.pdf

216. Mitler MM. Two-peak patterns in sleep, mortality and error. International Symposium on Sleep and Health Risk, Marburg/Lahn, Germany, Mar 1989.

217. Mitler MM, Carskadon MA, Czeisler CA, Dement WC, Dinges DF, Graeber RC. Catastrophes, sleep, and public policy: Consensus report. *Sleep*. 1988 Feb;11(1):100–109.

218. Mitler MM, Gujavarty K, Browman CP. Maintenance of wakefulness test: A polysomnographic technique for evaluating treatment efficacy in patients with excessive somnolence. *Electroencephalogr. Clin. Neurophysiol.* 1982 Jun;53(6):658–661.

219. Monk TH, Buysse DJ, Welsh DK, Kennedy KS, Rose LR. A sleep diary and questionnaire study of naturally short sleepers. *J. Sleep Res.* 2001 Sep;10(3):173–179.

220. Morgan DR. *Sleep Secrets for Shiftworkers & People with Off-beat*

Schedules. 1st ed. Whole Person Associates; 1996.

221. Muktār G al-D. *General History of Africa II: Ancient Civilizations of Africa*. London; Berkeley CA; Paris: Currey☐; University of California Press☐; Unesco; 1990.

222. Multiple Authors. Watch system. Wikipedia Free Encycl. 2014 May 5. Available from: http://en.wikipedia.org/w/index.php?title=Watch_system&oldid=601189355

223. Nah WL. *The Mathematics of the Longitude*. 2001. Available from: http://www.math.nus.edu.sg/aslaksen/projects/wln.pdf

224. Naitoh P, Banta GR, Kelly TL, Adrien J, Burr R. *Sleep Logs: Measurement of Individual and Operational Efficiency*. San Diego CA: Naval Health Research Center; 1991. Available from: http://www.dtic.mil/docs/citations/ADA239774

225. Naitoh P, Beare A, Biersner R. *Altered Circadian Periodicities in Oral Temperature and Mood in Men on an 18-Hour Work-Rest Cycle during a Nuclear Submarine Patrol*. San Diego CA: Naval Health Research Center; 1983. Available from: http://www.ntis.gov/search/product.aspx?ABBR=ADA102590

226. Nash N. Alternative watchkeeping. *Seaw. Int. J. Naut. Inst*. 2010 Nov;3–5.

227. National Research Council (U.S.), Committee on the Effect of Smaller Crews on Maritime Safety, National Research Council (U.S.), Commission on Engineering and Technical Systems. *Crew Size and Maritime Safety*. Washington, D.C.: National Academy Press; 1990.

228. National Research Council (U.S.), Committee on the Effects of Commuting on Pilot Fatigue. *The Effects of Commuting On Pilot Fatigue*. Washington, D.C.: National Academies Press; 2011.

229. National Transportation Safety Board. *Grounding of the U.S. Tankship Exxon Valdez on Bligh Reef, Prince William Sound, near Valdez, Alaska, March 2, 1989*. Washington DC: National Transportation Safety Board; 1990.

230. Nelson-Wong E, Callaghan JP. The impact of a sloped surface on

low back pain during prolonged standing work: a biomechanical analysis. *Appl. Ergon.* 2010 Oct;41(6):787–795.

231. Nelson-Wong E, Gregory DE, Winter DA, Callaghan JP. Gluteus medius muscle activation patterns as a predictor of low back pain during standing. *Clin. Biomech. Bristol Avon.* 2008 Jun;23(5):545–553.

232. Neri DF, Dinges DF, Rosekind MR. *Sustained Carrier Operations: Sleep Loss, Performance, and Fatigue Countermeasures.* NASA report. Moffett Field CA: NASA Ames Research Center; 1997.

233. Ngomo S, Messing K, Perrault H, Comtois A. Orthostatic symptoms, blood pressure and working postures of factory and service workers over an observed workday. *Appl. Ergon.* 2008 Nov;39(6):729–736.

234. Nguyen JL. *The Effects of Reversing Sleep-Wake Cycles on Sleep and Fatigue on the Crew of USS John C. Stennis.* Monterey CA: Naval Postgraduate School; 2002. Available from: http://stinet.dtic.mil/oai/oai?&verb=getRecord&metadataPrefix=html&identifier=ADA407035

235. Nicolas NH. *A History of the Royal Navy, from the Earliest Times to the Wars of the French Revolution.* London: Richard Bentley; 1847. Available from: http://books.google.com/books?id=rh8EAAAAQAAJ&printsec=frontcover&source=gbs_ge_summary_r&cad=0#v=onepage&q&f=false

236. Nilsson MP. *Primitive Time-Reckoning: A Study in the Origins and First Development of the Art of Counting Time among the Primitive and Early Culture Peoples.* Lund, Sweden: CWK Gleerup; 1920. Available from: http://www.abebooks.com/servlet/BookDetailsPL?bi=10165829479&searchurl=an%3Dnilsson%2Bmartin%2Bp%2Bfielden%2Bf%2Bj%26amp%3Bbsi%3D0%26amp%3Bds%3D30

237. Office of Technology Assessment. *Biological Rhythms: Implications for the Worker. New Developments in Neuroscience.* Washington, DC: U.S. Congress, Office of Technology Assessment, U.S. Government Printing Office. 1991. Available from: http://www.ntis.gov/search/product.aspx?ABBR=PB92117589

238. O'Hanlon JF, McCauley ME. *Motion Sickness Incidence as a Function of*

the *Frequency and Acceleration of Vertical Sinusoidal Motion*. Goleta CA: Human Factors Research Div, Canyon Research Group Inc. 1973. Available from: http://www.dtic.mil/docs/citations/AD0768215

239. O'Hanlon JF, Miller JC, Royal JW. *Effects of Simulated Surface Effect Ship Motions on Crew Habitability. Phase II. Volume 4*. Crew Cognitive Functions, Physiological Stress, and Sleep. Goleta CA: Canyon Research Group Inc, Human Factors Research Div; 1977. Available from: http://www.dtic.mil/docs/citations/ADA102830

240. Oldenburg M, Baur X, Schlaich C. Occupational risks and challenges of seafaring. *J. Occup. Health*. 2010;52(5):249–256.

241. Oldenburg M, Hogan B, Jensen H-J. Systematic review of maritime field studies about stress and strain in seafaring. *Int. Arch. Occup. Environ. Health*. 2013 Jan;86(1):1–15.

242. D' Onofrio P, Petersen H, Schwarz J, Pantaleev B, Sharma D, Åkerstedt T. Physiological sleepiness during a 7 day simulated sea voyage using the 6 h on/6 h off watch system. 21st International Symposium on Shift work and Working Time: Biological mechanisms, recovery and risk management in the 24 h Society, Stockholm, 28 Jul 2011.

243. Osborn C. *An Analysis of the Effectiveness of a New Watchstanding Schedule for U.S. Submariners*. Monterey CA: Naval Postgraduate School; 2004. Available from: http://www.dtic.mil/srch/doc?collection=t3&id=ADA427686

244. Paul MA, Ebisuzaki D, McHarg J, Hursh SR, Miller JC. *An Assessment of Some Watch Schedule Variants Used in Cdn Patrol Frigates: OP Nanook 2011*. 2012. Toronto, Canada: Defence Research and Development Centre. Available from: http://oai.dtic.mil/oai/oai?verb=getRecord&metadataPrefix=html&identifier=ADA568627

245. Paul MA, Gray GW, Lieberman HR, Love RJ, Miller JC, Trouborst M, et al. Phase advance with separate and combined melatonin and light treatment. *Psychopharmacology* (Berl.). 2011 Mar;214(2):515–523.

246. Paul MA, Gray GW, Nesthus TE, Miller JC. *An Assessment of the CF Submarine Watch Schedule Variants for Impact on Crew Performance*. Toronto, Canada: Defence Research and Development Centre;

2008. Available from: http://pubs.drdc-rddc.gc.ca/BASIS/pcandid/www/engpub/DDW?W%3DCA_RPNUM++%3D+%27DRDC-TORONTO-TR-2008-007%27%26M%3D1%26K%3D529839%26U%3D1

247. Paul MA, Gray GW, Nesthus TE, Miller JC. *An Assessment of the CF Submarine Watch Schedule Variants for Impact on Modeled Crew Performance.* Toronto, Canada: Defence Research and Development Centre; 2008. Available from: http://www.dtic.mil/srch/doc?collection=t3&id=ADA485455

248. Paul MA, Hursh SR, Miller JC. *Alternative Submarine Watch Systems: Recommendation for a New CF Submarine Watch Schedule.* Toronto, Canada: Defence Research and Development Centre; 2010.

249. Paul MA, Hursh SR, Miller JC. *Alternative Submarine Watch Systems: Recommendation for a New CF Submarine Watch Schedule.* Toronto, Canada: Defence Research and Development Centre; 2010. Available from: http://www.dtic.mil/srch/doc?collection=t3&id=ADA517285

250. Paul MA, Miller JC, Gray GW, Love RJ, Lieberman HR, Arendt J. Melatonin treatment for eastward and westward travel preparation. *Psychopharmacology (Berl.).* 2010 Feb;208(3):377–386.

251. Pettin TJ. *Fatigue As the Cause Of Marine Accidents, 1981–1985.* U.S. Coast Guard Marine Investigation Division; 1987.

252. Phang SE. *Roman Military Service: Ideologies of Discipline in the Late Republic and Early Principate.* Cambridge University Press; 2008.

253. Phillips R. Sleep, watchkeeping and accidents: a content analysis of incident at sea reports. *Transp. Res. Part F, Traffic Psychol. Behav.* 2000 Dec;3(4):229–240.

254. Phillips RO. *An Assessment of Studies of Human Fatigue in Land and Sea Transport.* Report no. 1354/2014. Oslo, Norway: Institute of Transport Economics, Dec 2014. Available from: http://trid.trb.org/view.aspx?id=1344403

255. Pigeau, Naitoh, Buguet, McCann, Baranski, Taylor, et al. Modafinil, d-amphetamine and placebo during 64 hours of sustained mental work. I. Effects on mood, fatigue, cognitive performance and body temperature. *J. Sleep Res.* 1995 Dec;4(4):212–228.

256. Plett R, Colquhoun WP, Condon R, Knauth P, Rutenfranz J, Eickhoff S. Work at sea: a study of sleep, and of circadian rhythms in physiological and psychological functions, in watchkeepers on merchant vessels. III. Rhythms in physiological functions. *Int. Arch. Occup. Environ. Health.* 1988;60(6):395–403.

257. Pollard JK, Sussman ED, Stearns M. *Shipboard Crew Fatigue, Safety and Reduced Manning.* Washington DC: U.S. Maritime Administration; 1990. Available from: http://ntl.bts.gov/lib/33000/33400/33413/33413.pdf

258. Post WM, Langefeld JJ. *Safety Consequences Onboard Shortsea Ships Due to A New Way of Working.* Soesterberg, The Netherlands: The Netherlands Organisation for Applied Scientific Research (TNO); 2011.

259. Post W, Punte P, Rasker P, Langefeld A. *Ship Manning Innovation: An Onboard Instrument for Measuring Safe Sailing.* IMO Resolut. A. 2003;890:21.

260. Putilov AA, Donskaya OG. Construction and validation of the EEG analogues of the Karolinska sleepiness scale based on the Karolinska drowsiness test. *Clin. Neurophysiol.* 2013 Jul;124(7):1346–1352.

261. Raby M, Lee JD. Fatigue and workload in the maritime industry. In: Hancock PA, Desmond PA, (ed.). *Stress, Workload and Fatigue.* CRC Press; 2000. p. 566–578.

262. Raby M, McCallum MC. Procedures for Investigating and Reporting Fatigue Contributions to Marine Casualties. *Proc. Hum. Factors Ergon. Soc. Annu. Meet.* 1997 Oct 1;41(2):988–992.

263. Rawlinson G. Phoenicia, Phoenician Ships, Navigation and Commerce. In: *History of Phoenicia.* London: Longmans, Green; 1889. Available from: http://phoenicia.org/ships.html

264. Rechtschaffen A, Kales A. *A Manual of Standardized Terminology, Techniques and Scoring System for Sleep Stages of Human Subjects.* Washington DC: National Institutes of Health; 1968.

265. Reeves DL, Winter KP, Bleiberg J, Kane RL. ANAM® Genogram: Historical perspectives, description, and current endeavors. *Arch. Clin. Neuropsychol.* 2007 Feb;22, Supplement 1:15–37.

266. Richards EG. *Mapping Time: The Calendar and Its History*. Oxford: Oxford University Press; 1999.

267. Richards ER, O'Brien BJ. *Misreading Scripture with Western Eyes: Removing Cultural Blinders To Better Understand The Bible*. Downers Grove ILl: IVP Books; 2012.

268. Roberts DA. *Analysis of Alternative Watch Schedules for Shipboard Operations: A Guide for Commanders*. Monterey CA: Naval Postgraduate School; 2012. Available from: http://www.dtic.mil/docs/citations/ADA561219

269. Ross JM. *Human Factors for Naval Marine Vehicle Design and Operation*. Surrey, England: Ashgate Publishing; 2009.

270. Rousmaniere J. The Art and Science of Standing Watch. *SailNet.com*. 2004 Feb 8. Available from: http://sailnet.com/forums/cruising-articles/20375-art-science-standing-watch.html

271. Rupp TL, Wesensten NJ, Bliese PD, Balkin TJ. Banking sleep: realization of benefits during subsequent sleep restriction and recovery. *Sleep*. 2009;32(3):311–321.

272. Russo M, Thomas M, Thorne D, Sing H, Redmond D, Rowland L, et al. Oculomotor impairment during chronic partial sleep deprivation. *Clin. Neurophysiol*. 2003 Apr;114(4):723–736.

273. Rutenfranz J, Aschoff J, Mann H. The effects of a cumulative sleep deficit, duration of preceding sleep period and body temperature on multiple choice reaction time. In: Colquhoun WP, editor. *Aspects of Human Efficiency: Diurnal Rhythm and Loss of Sleep*. London: English Univesities Press; 1972. p. 217–230.

274. Rutenfranz J, Plett R, Knauth P, Condon R, De Vol D, Fletcher N, et al. Work at sea: a study of sleep, and of circadian rhythms in physiological and psychological functions, in watchkeepers on merchant vessels. II. Sleep duration, and subjective ratings of sleep quality. *Int. Arch. Occup. Environ. Health*. 1988;60(5):331–339.

275. Sanquist TE, Raby M, Maloney A, Carvalhais A. *Fatigue And Alertness In Merchant Marine Personnel: A Field Study Of Work And Sleep Patterns*. Seattle WA: Battelle Research Center; 1996. Available from: http://www.dtic.mil/srch/doc?collection=t3&id=ADA322126

276. Sanquist TF, Raby M, Forsythe A, Carvalhais A. Fatigue in Merchant Marine Personnel. *Proc. Hum. Factors Ergon. Soc. Annu. Meet.* 1997 Oct 1;41(2):983–987.

277. Sasseville A, Paquet N, Sévigny J, Hébert M. Blue blocker glasses impede the capacity of bright light to suppress melatonin production. *J. Pineal Res.* 2006 Aug;41(1):73–78.

278. Sawyer TL. *The Effects of Reversing Sleep-Wake Cycles on Mood States, Sleep, and Fatigue on the Crew of the USS JOHN C. STENNIS*. Monterey CA: Naval Postgraduate School; 2004. Available from: http://www.dtic.mil/srch/doc?collection=t3&id=ADA424687

279. Schaefer K, Kerr C, Buss D, Haus E. Effect of 18-h watch schedules on circadian cycles of physiological functions during submarine patrols. *Undersea Biomed. Res.* 1979;6 Suppl:S81–90.

280. Scott A, LaDou J. Shiftwork: effects on sleep and health with recommendations for medical surveillance and screening. *Occup. Med. Phila. PA*. 1990 Jun;5(2):273–299.

281. Shattuck NL, Matsangas P. *Work and Rest Patterns and Psychomotor Vigilance Performance of Crewmembers of the USS Jason Dunham: A Comparison of the 3/9 And 6/6 Watchstanding Schedules*. Monterey CA: Naval Postgraduate School; 2014.

282. Shattuck NL, Matsangas P. A six-month assessment of sleep during naval deployment: A case study of a commanding officer. *Aviat. Med. Hum. Perform.* 2015;accepted.

283. Shattuck NL, Matsangas P, Brown S. *A Comparison Between the 3/9 and the 5/10 Watchbills*. Monterey CA, Naval Postgraduate School; 2015. Available from: https://calhoun.nps.edu/handle/10945/45008

284. Shattuck NL, Matsangas P, Moore J, Wegemann L. *Prevalence of musculoskeletal symptoms, excessive daytime sleepiness, and fatigue in the crew memebers of a U.S. Navy ship*. Monterey CA: Naval Postgraduate School; 2015. Available from: https://calhoun.nps.edu/handle/10945/44989

285. Shattuck NL, Matsangas P, Powley EH. *Sleep Patterns, Mood, Psychomotor Vigilance Performance, and Command Resilience of Watchstanders on The "Five And Dime" Watchbill*. Monterey CA: Naval

Postgraduate School; 2015.

286. Shattuck NL, Matsangas P, Waggoner L. Assessment of a Novel Watchstanding Schedule on an Operational us Navy Vessel. *Proc. Hum. Factors Ergon. Soc. Annu. Meet.* 2014 Sep 1;58(1):2265–2269.

287. Shattuck NL, Waggoner LB, Young RL, Smith CS, Brown SAT, Matsangas P. Shiftwork Practices in the United States Navy: A Study of Sleep and Performance in Watchstanders aboard the USS Jason Dunham. 21st International Symposium on Shiftwork and Working Time, Costa do Sauipe, Brazil: Nov 2013.

288. Shay J. Ethical standing for commander self-care: the need for sleep. *Parameters* (US Army War Coll. Press). 1998;Summer:93–105.

289. Shilling CW, Kohl JW. *History of Submarine Medicine in World War II.* New London CT: U.S. Naval Submarine Base; 1947. Available from: https://ia801702.us.archive.org/33/items/SubmarineMedicineInWorldWarII/Submarine%20Medicine%20in%20World%20War%20II.pdf

290. Skuld Assurance Society. *How to Prevent and Mitigate Fatigue.* 2007 May;

291. Sluiter JK, van der Beek AJ, Frings-Dresen MH. The influence of work characteristics on the need for recovery and experienced health: a study on coach drivers. *Ergonomics.* 1999;42(4):573–583.

292. Smith A. *Clocks and Watches.* Rev Edition. Ebury Press; 1989.

293. Smith A. *Adequate Crewing and Seafarers' Fatigue: The International Perspective.* Cardiff, Wales: Centre for Occupational and Health Psychology, Cardiff University; 2007.

294. Smith A, Allen P, Wadsworth E. *Seafarer Fatigue: The Cardiff Research Programme.* Cardiff, Wales: Centre for Occupational Health and Psychology, Cardiff University; 2006.

295. Sobel D. *Longitude: The True Story of A Lone Genius Who Solved the Greatest Scientific Problem of His Time.* New York: Penguin; 1996.

296. Sood A. Medical screening and surveillance of shift workers for health problems. *Clin. Occup. Environ. Med.* 2003 May;3(2):339–349.

297. Stern JA, Walrath LC, Goldstein R. The endogenous eyeblink.

Psychophysiology. 1984 Jan;21(1):22–33.

298. Stimson A. The Mariner's Astrolabe: A Survey of Known, Surviving Sea Astrolabes. *UC Biblioteca Geral* 1; 1988.

299. Stolgitis W. *The Effects of Sleep Loss and Demanding Work/Rest Cycles: An Analysis of the Traditional Navy Watch System and a Proposed Alternative.* Monterey CA: Naval Postgraduate School; 1969. Available from: http://stinet.dtic.mil/oai/oai?&verb=getRecord&metadataPrefix=html&identifier=AD0706027

300. Strong R, Brown D. *Task Performance and Sleep in the Two Watch System: Report of a Trial in a Royal Naval Fleet Patrol Submarine (SSN).* Alverstoke UK: Institute of Naval Medicine; 1989.

301. Szpakowska KM. *Behind Closed Eyes: Dreams and Nightmares in Ancient Egypt.* Swansea, Wales; Oakville, CT: Classical Press of Wales. 2003.

302. Thorne H, Hampton S, Morgan L, Skene D, Arendt J. Differences in sleep, light, and circadian phase in offshore 18:00-06:00 h and 19:00-07:00 h shift workers. *Chronobiol. Int.* 2008 Apr;25(2):225–235.

303. Torsvall L, Castenfors K, Åkerstedt T, Fröberg J. Sleep at sea: A diary study of the effects of unattended machinery space watch duty. *Ergonomics.* 1987;30(9):1335–1340.

304. Trousselard M, Leger D, van Beers P, Coste O, Vicard A, Pontis J, et al. Sleeping under the Ocean: Despite Total Isolation, Nuclear Submariners Maintain Their Sleep and Wake Patterns throughout Their Under Sea Mission. *PLOS ONE.* 2015 May 27;10(5):e0126721.

305. Truxtum T. *Short Account of the Several General Duties of Officers, Of Ships of War.* Navy Dep. Libr. Nav. Hist. Herit. Command. 1794. Available from: http://www.history.navy.mil/faqs/faq59-2.htm

306. Turner HR. *Science in Medieval Islam: An Illustrated Introduction.* Austin TX: Univesity of Texas Press; 1997. Available from: http://utpress.utexas.edu/index.php/books/tursci

307. United Kingdom P&I Club. *Analysis of Major Claims.* The United Kingdom Mutual Steam Ship Assurance Association (Bermuda) Limited, Thomas R. Miller & Son Inc.; 1993.

308. United States Coast Guard. *Report on Demonstration Project: Implementing the Crew Endurance Management System (CEMS) on Towing Vessels*. Washington DC: Department of Homeland Security; 2005.

309. U.S. Fleet Forces Command. *Naval Customs, Traditions, & Etiquette*. Nav. Cust. Tradit. Etiquette. 2014. Available from: http://www.public.navy.mil/usff/Pages/customs.aspx#dogwatch

310. U.S. Naval History and Heritage Command. *Bells on Ships*. 2014. Available from: http://www.history.navy.mil/faqs/faq83-1.htm

311. U.S. Naval History and Heritage Command. *Rules for the Regulation of the Navy of the United Colonies of North-America*. 1775 Nov 28. Available from: http://www.history.navy.mil/faqs/faq59-5.htm

312. Utterback R, Ludwig G. *A Comparative Study of Schedules for Standing Watches Aboard Submarines Based on Body Temperature Cycles*. Bethesda MD: Naval Medical Research Inst.; 1949. Available from: http://www.ntis.gov/search/product.aspx?ABBR=AD667707

313. Virk Z. Transmission of Islamic science to Europe & Renaissance. *Muslim Times*. 2014 Aug 1. Available from: http://www.themuslimtimes.org/2013/03/europe-and-australia/europe/transmission-of-islamic-science-to-europe-renaissance

314. Virk Z. *Science and Technology in Islamic Spain*. Toronto, Canada: Available from: https://www.academia.edu/6431626/Science_and_Technology_in_Islamic_Spain

315. Wadsworth EJK, Allen PH, McNamara RL, Smith AP. Fatigue and health in a seafaring population. *Occup. Med. Oxf. Engl.* 2008 May;58(3):198–204.

316. Wall JT. *Crossing Old Trails to New in North Central Wyoming*. Philadelphia: Dorrance; 1973.

317. Warm JS, Parasuraman R, Matthews G. Vigilance Requires Hard Mental Work and Is Stressful. *Hum. Factors*. 2008 Jun 1;50(3):433–441.

318. Waterhouse J, Reilly T, Atkinson G, Edwards B. Jet lag: trends and coping strategies. *Lancet*. 2007 Mar 31;369(9567):1117–1129.

319. Weibel L, Follenius M, Spiegel K, Gronfier C, Brandenberger G. Growth hormone secretion in night workers. *Chronobiol. Int.* 1997 Jan;14(1):49–60.

320. Wever R. Characteristics of circadian rhythms in human functions. *J. Neural Transm.* Suppl. 1986;21:323–373.

321. Wever R. *The Circadian System of Man.* New York: Springer-Verlag; 1979.

322. Wever R, Poláscaronek J, Wildgruber C. Bright light affects human circadian rhythms. *Pflüg. Arch. Eur. J. Physiol.* 1983 Jan;396(1):85–87.

323. Wilkinson R, Edwards R. *Stable Hours and Varied Work as Aids to Efficiency.* Royal Naval Personnel Research Committee, Medical Research Council; 1969. Available from: http://www.dtic.mil/docs/citations/AD0734527

324. Wilkinson RT. The effect of lack of sleep on visual watch-keeping. *Q. J. Exp. Psychol.* 1960;12(1):36–40.

325. Wilkinson RT, Edwards RS. Stable hours and varied work as aids to efficiency. *Psychon. Sci.* 1968;13:205–206.

326. Wilkinson RT, Houghton D. Field test of arousal: a portable reaction timer with data storage. *Hum. Factors.* 1982 Aug;24(4):487–493.

327. Williams RL, Karacan I, Hursch CJ. *Electroencephalography (EEG) of Human Sleep: Clinical Applications.* New York: Wiley; 1974. Available from: http://www.getcited.org/pub/101416203

328. Woodward DP, Nelson PD. *A User Oriented Review of the Literature on the Effects of Sleep Loss, Work-Rest Schedules, and Recovery on Performance.* Arlington VA: Office of Naval Research, 1974. Available from: http://www.dtic.mil/dtic/tr/fulltext/u2/a009778.pdf

329. Yokeley MT. *Effects of Sleep Deprivation on U.S. Navy Surface Ship Watchstander Performance using Alternative Watch Schedules.* Monterey CA: Naval Postgraduate School; 2012. Available from: http://www.dtic.mil/docs/citations/ADA567421

330. Young CR, Jones GE, Figueiro MG, Soutière SE, Keller MW, Richardson AM, et al. At-Sea Trial of 24-h-Based Submarine Watchstanding Schedules with High and Low Correlated Color

Temperature Light Sources. *J. Biol. Rhythms.* 2015 Apr;30(2):144–154.

331. Young RL. *A Comparison of Sleep and Performance of Sailors on an Operationally Deployed U.S. Navy Warship.* Monterey CA: Naval Postgraduate School; 2013. Available from: http://www.dtic.mil/docs/citations/ADA589578

Appendix
Watch Plan Analyses and Detailed Analysis Results

The appendix describes the methods I used to analyze the nominal watch plans shown in this book. I used the Sleep, Activity, Fatigue, and Task Effectiveness (SAFTE) model and its software implementation, The Fatigue Avoidance Scheduling Tool (FAST) to model the effects of each watchstanding plan shown in this book quantitatively (48, 136–141, 198). Of these various metrics, I reported in the text of the book those that might be of interest to the reader. A complete set of metrics is available in the full Appendix. The Appendix is available from the author, and I have also posted it on ResearchGate.

About the Author

I provide consulting services in Fatigue Risk Management Systems (FRMS) based upon over 45 years of applied research and development concerning human cognitive performance and fatigue. I've focused mainly on the measurement and analysis of human physical and cognitive performance in military and civil aviation; highway, rail and maritime transportation; and night and shift work. Operator fatigue has been at the center of my interests since my days as an Air Force pilot in the C-130E Hercules tactical transport in Vietnam. In addition to this series on shiftwork, I'm also the author of *Fatigue* in McGraw-Hill's Controlling Pilot Error series (2001), the ASIS CRISP report *Fatigue Effects and Countermeasures in 24/7 Security Operations* (2010), "An historical view of operator fatigue" in *The Handbook of Operator Fatigue* (Ashgate, 2012, Chapter 2), and *Cognitive Performance Research at Brooks Air Force Base, Texas, 1960-2009* (Smashwords, 2013). My specialties include:

- Fatigue effects on worker productivity and risk during 24/7 work
- Fatigue effects on driver performance in 24/7 transportation operations
- Shiftwork schedule analysis and improvement
- Psychological and environmental effects on human physiology
- Fatigue associated with circadian rhythm effects
- Forensic investigations of accidents caused by fatigue, poor vigilance, lack of sleep and night work

www.ingramcontent.com/pod-product-compliance
Lightning Source LLC
Chambersburg PA
CBHW071017240526
45469CB00006BD/1962